Measuring Stress in Humans

The purpose of this book is to present state-of-the-art non-invasive methods of measuring the biological responses to psychosocial stress in non-laboratory (field) settings. Following the pathways of Seyle's General Adaptation Syndrome, the text first describes how to assess the psychosocial stressors of everyday life and then outline how to measure the psychological, behavioral, neurohumeral, physiological and immunological responses to them. The book concludes with practical information on assessing special populations, analyzing the often complicated data that are collected in field stress studies and the ethical treatment of human subjects in stress studies. It is intended to be a practical guide for developing and conducting psychophysiological stress research in human biology. This book will assist students and professionals in designing field studies of stress.

Gillian H. Ice is Associate Professor in the Department of Social Medicine at Ohio University of Osteopathic Medicine.

Gary D. James is Director of the Institute for Primary and Preventative Health Care, and Professor of Nursing and Anthropology at Binghamton University.

Cambridge Studies in Biological and Evolutionary Anthropology

Series Editors

HUMAN ECOLOGY
C.G. Nicholas Mascie-Taylor, University of Cambridge
Michael A. Little, State University of New York, Binghamton
GENETICS
Kenneth M. Weiss, Pennsylvania State University
HUMAN EVOLUTION
Robert A. Foley, University of Cambridge
Nina G. Jablonski, California Academy of Science
PRIMATOLOGY
Karen B. Strier, University of Wisconsin, Madison

Measuring Stress in Humans

A Practical Guide for the Field

EDITED BY

GILLIAN H. ICE
Department of Social Medicine
Ohio University of Osteopathic Medicine

and

GARY D. JAMES
Institute for Primary and Preventative Health Care and
Decker School of Nursing
Binghamton University

CAMBRIDGE
UNIVERSITY PRESS

CAMBRIDGE UNIVERSITY PRESS
Cambridge, New York, Melbourne, Madrid, Cape Town,
Singapore, São Paulo

Cambridge University Press
The Edinburgh Building, Cambridge CB2 2RU, UK
Published in the United States of America by Cambridge
University Press, New York

www.cambridge.org
Information on this title: www.cambridge.org/9780521844796

First published 2007

Printed in the United Kingdom at the University Press, Cambridge

A catalogue record for this publication is available from the British Library

Library of Congress Cataloging in Publication data

ISBN-13 978-0-521-84479-6 hardback
ISBN-10 0-521-84479-7 hardback

Contents

Contributors

Dan E. Brown
Department of Anthropology, University of Hawaii at Hilo, 200 W
Kawili Street, Hilo, HI 96720-4091

William W. Dressler
Department of Anthropology, PO Box 870210, University of Alabama,
Tuscaloosa, AL 35487-0210

Gillian H. Ice
Assistant Professor, Department of Social Medicine and Director of
International Programs, Ohio University College of Osteopathic
Medicine, 309 Grosvenor Hall, Athens, OH 45701

Gary D. James
Decker School of Nursing, Binghamton University, SUNY, Academic B,
Room 326, Binghamton, NY 13902-6000

Tessa M. Pollard
University of Durham, 43 Old Elvet, Durham DH1 3HN, UK

Thomas W. McDade
Department of Anthropology, Northwestern University, 1810 Hinman
Avenue, Evanston, IL 60202

Sharon R. Williams
2711 W Ridgeland Avenue, Waukegan, IL 60085

Foreword

Stress has been recognized as an important psycho-physiological state since the pioneering work of Hans Selye. But until quite recently it has mainly been perceived in humans as a condition generated by extreme and hostile environments such as going into battle, hospital or academic examinations. Increasingly, however, it has been identified as being a consequence of many aspects of lifestyle and the events of everyday living and that, to varying degrees, large numbers of people experience it. Indeed, from the point of view of long-term health, low-level frequent chronic stress is likely to be much more important than occasional acute episodes.

Chronic stress can hardly be studied by experimental procedures in the laboratory. It clearly needs a population approach with investigators monitoring people in the "field" as they go about their daily business. Psychologists have gained important insights by the design of questionnaires which can be applied not only to particular groups undertaking activities which are deemed to be stressful, such as air traffic controllers, but also to whole populations, experiencing a diversity of lifestyles. They have identified various elements, particularly in occupational situations, which aggravate stress, as for example absence of job control, but questionnaires are of little use outside one's own language, or at least culture. They also have questionable validity in the study of children.

For these wider studies it is necessary, or at least desirable, to have some physiological measures of the stressed state, either of the homeostatic mechanisms which are elicited to control stress or of the morbid consequences of being stressed. In principle, such measures are not only objective but also appropriate for any population or population group situation in any culture. They also avoid the dangerous pre-judgment of whether or not some environment is stressful, for what can generate great stress in one individual may cause none in another. Environments may certainly contain stressors, but stress itself is a phenomenon of the organism not of the environment. Having said that, there are many problems both theoretical and practical in both obtaining the desired physiological information and interpreting it. Some, such as blood pressure, can usually be obtained fairly easily, though readings can be very labile. However, others, such as hormone levels in urine or saliva, can be immensely difficult. For many purposes, especially cross-cultural comparisons, one would like 24-hour urine samples, but even with the most willing and co-operative of subjects, one or more urinations are likely to be missed, unless the subject is confined to a hospital bed!

Then there are the complex problems of interpretation. Epinephrine, for example, is often referred to as a stress hormone. The excretion is certainly greatly raised when people go into battle, examinations or competitive sport, but it is also raised in those playing in a pop band and lowered in those who report being endlessly bored. It would seem to reflect levels of psychological arousal and while many unpleasant experiences cause arousal, so can those we enjoy. Few would call a good party stressful.

Matters such as these are discussed at great length and with great authority in this book. It covers all the physiological approaches to measuring stress, considering both broad theoretical issues and the practical methods that have been used. It sets these discussions within the wider framework of study design, varying culture and research ethics with populations. It is surely indispensable for any anthropologist studying stress, but with its emphasis on practical matters it should also be of great value to clinicians, psychologists and physiologists. It has no competitors; there is no other book like it.

Geoffrey A. Harrison

Part I
General principles

1 *Conducting a field study of stress: general principles*

GILLIAN H. ICE AND GARY D. JAMES

Introduction

In recent years, interest in the study of stress has expanded, particularly in the disciplines of anthropology and human biology where the focus of research has drifted toward evaluating the adaptive biological, cultural and psychological responses to stressors inherent in everyday living. Field studies of stress in these disciplines have been conducted across a wide continuum of contexts. These range from an assessment of the stress of "modernization" where biological and cultural responses of populations undergoing rapid industrialization/Westernization are examined, e.g. James *et al.* (1985), to the responses of people facing novel, new environments in modern, Western societies, such as occur in nursing homes (Ice *et al.*, 2002).

Biologically, in studying the stress response, measurements can be made at any of several junctures in the psychophysiological pathways between stressful stimuli and adaptive or pathological outcomes. Measures include emotional/behavioral responses, hormonal variation in the sympathetic adrenal medullary system (SAMS), hormonal variation in the hypothalamic pituitary adrenal axis (HPA), physiological changes in the cardiovascular system and enhanced immune responses. The choice of the most appropriate measure will be determined by a combination of the goals of research, the population of interest and practical concerns. In addition to the particular stress marker(s) chosen for a study, the appropriate sampling strategy and design must be determined. These range from group comparisons of a one-time measure, to ecological momentary assessments, to multiple measures evaluated in a longitudinal design. Some study designs used in stress research require special analytic strategies and sophisticated statistical approaches.

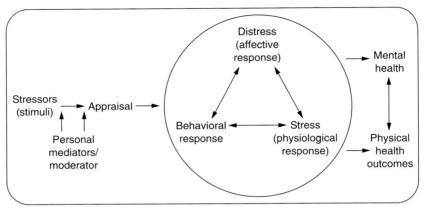

Figure 1.1.   The stress process.

While stress markers themselves are common outcome measures, the ultimate goal in stress research is to determine the relationships between stress and health or stress and adaptation. To this end, several conceptual models have been proposed across a variety of disciplines. It is the intent of this chapter to first provide an overview of these models as well as the tools and instruments used to evaluate the stress experience by researchers in the various disciplines, and then to introduce an integrated model in which *stress is considered a process by which a stimulus elicits an emotional, behavioral and/or physiological response, which is conditioned by an individual's personal, biological and cultural context* (Figure 1.1).

Definitions of stress and their origins

Stress research can be confusing as there are a multitude of definitions which are often not equivalent. The term "stress" has been used to refer to at least three different components of the stress process: 1) the input or stimuli, 2) processing systems, including both physiological and psychological and 3) the output or stress response, e.g. rise in blood pressure (Mason, 1975; Levine and Ursin, 1991). Different disciplines have focused on different aspects of the stress process. Table 1.1 provides a general guide of different approaches by discipline.

Early research on the physiological processes related to stress has been described by Walter Cannon (Cannon, 1914), the author of the "flight or fight syndrome." He also coined the term "homeostasis" to describe the process of maintaining internal stability in the face of environmental change (Cannon, 1932). This term does not mean something fixed and unchanging, but a relatively constant, complex, well coordinated and usually stable condition. Cannon was also interested in determining the specific mechanisms of response to changes in the external environment, which allowed for optimal bodily function. He showed that there are specialized sensory nerves to communicate the state of the rest of the body to the brain, that the brain is able to detect non-optimal internal states, and that the brain can call a variety of mechanisms into play to compensate correctly. Finally, he noted that failure to maintain homeostasis could result in tissue damage or death, and he was among the first to examine the challenges of psychologically meaningful stimuli and the impact of moods (Cannon, 1929, 1932).

   Hans Selye popularized the concept of "stress" and many researchers trace the origin of its study and definition to Hans Selye and his 1936 paper, "A syndrome produced by diverse nocuous agents" (Selye, 1936). In this paper, he described stress as a non-specific response of the body to "noxious stimuli." Selye's concept described a physiological response to physical and physiological stimuli, described as stressors (Selye, 1946). Selye later named and elaborated on the process as the general adaptation syndrome (Selye, 1946). This syndrome has three stages: 1) alarm reaction, 2) stage of resistance and 3) stage of exhaustion. In the alarm stage, the body reacts to a stimulus by activating the hypothalamic pituitary adrenal (HPA) axis. The resistance stage signals successful adaptation to the stimulus. Exhaustion occurs when exposure to stimuli is prolonged. Selye believed that the body's stores of glucocorticoids (the output of the HPA axis) were depleted. Most researchers now believe that the body does not deplete stores of glucocorticoids but that prolonged exposure to a stressor results in suppression of the immune system and wear and tear of several body systems, which then places individuals at risk of a variety of disease outcomes.

   While Selye and biomedical researchers conducted their research on physical and physiological stressors in animal models, several investigators starting looking at the impact of psychosocial stressors in humans. The initial focus of these investigations was traumatic or major life events. Many credit Adolf Meyer and Harold Wolff in the 1930s and 1940s for early development of research examining stressful life events and illness (Rahe, 1989; Cohen *et al.*, 1997). Meyer suggested that physicians should record life events as part of their medical examination while Wolff went on to describe the association between life events and illness (Cohen *et al.*, 1997). One of the first published scales created to measure life events, "The Social Readjustment Rating Scale" (Holmes

Table 1.1. *Variables used by different disciplines*

Discipline	Definition of stress	Stressors	Processing/ appraisal	Interacting variables	Response	Outcomes
Biomedical animal research, including physiology, biological psychology & neurology	Non-specific response to stimulus	Cardiac catheterization; cold or heat exposure; competitive social interaction; food, sleeping or sensory deprivation; handling; isolation or crowding; immobilization	Rarely measured	Rarely measured	Hormonal and physiological	Organ damage
Social psychology	Transaction between the person and the environment	Daily hassles, life events	Focus of research	Demographics, personality factors, social resources	Perceived stress, physiological response	Mental health and physical health outcomes
Sociology	Psychosocial stress:	Chronic stressors (strains), role strain	Variably measured	Social values, social context,	Perceived stress, emotional,	Mental health outcomes,

	socially derived stressors which lead to negative affect or distress			social support, self-concept, life course	behavioral response	social relationship disruption
Cultural anthropology	Variable	Incongruity, cultural stressors	Frequently measured	Cultural context, resistance resources, social resources	Perceived stress, behavioral and physiological responses	Mental and physical health outcomes, cultural syndromes, local idioms of distress
Human biology/biological anthropology	Disruption in homeostasis or allostasis	Variable, including cultural, environmental and physical stressors	Rarely measured	Cultural context, life cycle	Physiological response	Physical health outcomes
Physiology, biomedical sciences, with a human focus	Allostasis	Variable, many human equivalents to animal stressors	Rarely measured	Demographic factors	Physiological response	Allostatic load, physical health outcomes

and Rahe, 1967) has become the foundation upon which most current life events scales are based. While these researchers were looking at humans, they were still working from a Selyen model of a non-specific response to stressors.

Stress research focused on major life events through the 1960s with researchers further expanding the kinds of events that might be considered stressful. As it started to become clear that there were individual differences in the response to such events, Lazarus and colleagues developed a theory of stress which emphasized appraisal and coping in the late 1960s and 1970s. Lazarus has argued in multiple publications that the "stimulus-centered perspective" of life events approach and the physiological approaches of Selye and Cannon were too simplistic (Lazarus, 1984; Lazarus and Folkman, 1984; Lazarus, 1999). Lazarus and colleagues suggested that the best way to view the stress process is as a transaction between the person and environment (Lazarus, 1984; Lazarus and Folkman, 1984; Lazarus, 1999). The impact of any potential stimulus is determined by an individual's appraisal and coping. Within this transaction, an individual goes through a cognitive assessment to determine if a particular circumstance is a threat and if s/he has the resources or coping skills to meet the demand placed upon him/her by the threat (Lazarus and Folkman, 1984). Lazarus and Folkman define coping as, "constantly changing cognitive and behavioral efforts to manage specific external and/or internal resources of the person" (Lazarus and Folkman, 1984, p. 141). In addition to emphasizing the importance of appraisal and coping Lazarus has suggested that we shift our focus from major life events to daily hassles (Lazarus, 1984, 1999). From this cognitive theoretical approach, there may be limitless types of transactions between the person and environment depending on the context and the person's age, culture and experience. While psychologists were expanding the model of stress, the Selyen model still had a great impact on future biomedical and neurological research (Elliot and Eisdorfer, 1982). As Selye maintained that the stress response was non-specific (i.e. did not vary by stressor), the stressors chosen by biomedical researchers using animal models were often based on convenience rather than research question (Elliot and Eisdorfer, 1982). The concept of a non-specific response has since been questioned as a number of researchers have demonstrated variation in response to different stressors (Mason, 1975). Mason (1975) cited several studies which demonstrate that the HPA axis response to stressors varies with the type of stressor and the experimental conditions. He further pointed out that many of the "physical stressors" used in animal experiments have a psychological component. For example, in his own research on starvation in monkeys, the HPA response was significantly diminished when monkeys were fed non-nutrient placebo food (Mason, 1974). The placebo acted to minimize the psychological effect of sudden deprivation. Mason was one of the first to cross the disciplinary boundaries by incorporating psychological models of stress in his biological models. Cassel (1976) further pointed out that the Selyen concept of a non-specific response led many to suggest that there is such a thing as a "stress state" or "stress disease." He suggested, as many have now come to believe, that stress does not produce a specific disease but rather places people in a state that makes them more susceptible to a range of diseases. He suggested that "psycho-social processes acting as 'conditional stressors' will, by altering the endocrine balance of the body, increase susceptibility of the organism to direct noxious stimuli, i.e. disease agents" (Cassel, 1976, p. 109). Both Mason and Cassel were influential in getting future researchers to think of stress as a process rather than a simple stimulus-response relationship suggested by biomedical researchers and those in the life-event arena.

Pearlin and colleagues (Dohrenwend and Pearlin, 1982; Pearlin, 1989) further criticized the life events and "stimulus−response" approaches on a number of accounts. First, they questioned the theory that all change is harmful and suggested that only change that is undesirable, un-scheduled, non-normative and uncontrollable is harmful. Second, they suggested that the life-events approach treats events as if they occur in a vacuum without consideration of the socioeconomic context in which they occur. Further, they critiqued the instruments for measuring exposure to life events as a conflation of acute events with ongoing stressors. Pearlin suggested that health may be impacted not by the individual "major" event but by the "continuing circumstances in which the event is embedded" (Pearlin, 1989, p. 244). For example, if an individual forecloses on his/her house, s/he was likely to have experienced problems of continuous poverty and debt prior to the actual foreclosure. Pearlin did not suggest that life events are unimportant to the stress process, merely, that "some events under some conditions are powerful stressors that affect people's lives directly and indirectly" (Pearlin, 1989, p. 245). Pearlin emphasized the importance of looking at chronic stressors which he called strains. These were defined as "relatively enduring problems, conflicts and threats that many people face in their daily lives" (Pearlin, 1989, p. 245). His research particularly emphasized the importance of role strains which are problems connected to the social roles which people fill. Most importantly, Pearlin suggested that we should not look at chronic stressors or major life events in a vacuum as they may come together to produce stress in a number of ways (Pearlin, 1983). Life events may lead to chronic stressors or result from chronic stressors and they can interact and provide meaning for one another. While Lazarus focused on cognitive appraisal as a mediator of the stress process, Pearlin discussed the importance of social values in mediating the impact of a stressor. These social values "regulate the effects of experience by regulating the meaning and importance of the experience" (Pearlin, 1989, p. 249). ✗

Lazarus, who emphasizes the individualistic aspect of the stress process, cautions against the socio-cultural approach to the stress process in which the impact of social structure or culture results in or mediates stress (Lazarus, 1999). By examining the stress process in this manner, Lazarus and Folkman argue that generalizations based on such analyses simplify our understanding of the process and distill a dynamic process into a static one in which people are treated as "carbon copies" as opposed to individual cognitive and emotional beings (Lazarus and Folkman, 1984). However, by examining social forces or social context, Pearlin is not really suggesting that individuals are carbon copies but that they do exist in a social context which should be considered when examining the stress process.

Dressler has been very influential in bringing cultural context into stress research (Dressler, 1991, 1995; Dressler and Bindon, 1997; Dressler and Bindon, 2000). In fact, he argues that culture has influence on multiple components of the stress process. Cultural context influences meaning of stressors, patterns of stressors and coping resources. Further, culture can be a stressor or it can be a mediator. His work on lifestyle incongruity and cultural consonance (further explained in Chapter 2) has had a tremendous impact on stress research within anthropology and human biology. These elegant models connect the individual process of stress with the social and cultural context, in a sense fusing the models of Lazarus and Pearlin.

Human Biologists (or Biological Anthropologists) have a history of looking at environmental stressors as a source of human variation. They often take the wider perspective that stressors are anything that take the body away from homeostasis and thus, by default, stress becomes a disruption in homeostasis. However, historically human biologists have approached stress in a very Selyen way. The traditional focus on adaptation to environmental stressors has carried over into the way that social stressors are often examined. Thus, human biologists have often started with a potential stressor and compared individuals exposed to those who are not. Often a "stress hormone" or health outcome is used to determine if the two groups differ in their level of stress. If there is a difference, this confirms that the potential stressor is indeed a stressor. Unfortunately, this research often left out notions of appraisal or socio-cultural context. With the maturing of stress research in human biology, models have become more complex and involve more of the stress process.

In 1988, Peter Sterling and Joseph Eyre introduced the concept of *allostasis,* literally "achieving stability through change" in order to provide a logical structure for understanding the ever-shifting integrated biobehavioral, endocrinological and physiological systems of the body that promote adaptation and drive natural selection (Sterling and Eyer, 1988). For example, were blood pressure to remain constant throughout the day, individuals would have difficulty responding to their own changing activities and other environmental variations. However, because it is part of an allostatic system, blood pressure will vary continuously to adapt the individual to the changing circumstances. Because it continuously changes, the individual does not have a single "homeostatic" blood pressure state per se, but rather has many stable states, which are directly related to the many and ever-changing internal and external environmental conditions to which the individual must adapt. The multiple stable states of blood pressure differentiate this physiological system from other bodily homeostatic systems such as those that maintain tissue pH. The HPA and SAMS axes also act as allostatic systems (McEwen and Stellar, 1993; McEwen, 1998a,b). Some have suggested that because the concept allostasis unifies the physiology of acute and long-term adaptations and stress responses as well as their outcomes into a single process, it should replace the stress concept (McEwen, 2002). As part of this argument, McEwen (McEwen and Stellar, 1993; McEwen, 1998a,b) has introduced the concept of allostatic load. While allostasis is critical to adaptation and survival, "allostatic load" is defined as "the price the body pays over long periods of time for adapting to challenges" (McEwen, 2001, p. 44). McEwen and colleagues have created an index of allostatic load and applied it to the health outcomes of participants in the MacArthur Aging Studies. To measure allostatic load, indicators of "system failure" (e.g. high blood pressure, large waist to hip ratios, elevated urinary epinephrine and cortisol, etc.) were tallied. The index was calculated from the number of indicators in which a participant's measurements fell in the uppermost (4th) quartile of the population distribution (Seeman *et al.*, 1997a). This index of allostatic load predicted declines in cognitive and physical functioning, cardiovascular disease and mortality (Seeman *et al.*, 1997a,b; Karlamangla *et al.*, 2002).

In many ways, the concept of allostasis can be seen as a new spin on the old Selyen concept of stress. However, recently Schulkin (2004) edited a volume entitled *Allostasis, Homeostasis, and the Costs of Physiological Adaptation* in which a spectrum of physiological and biobehavioral processes were recast and evaluated from the perspective of allostasis. Based on the discussions in this volume, it is quite possible that the concept of allostasis may join homeostasis as the foundation for future understanding of the relationship between stress and adaptation.

Putting the models together: the stress process

While the stress field has matured, it is still marked with disciplinary differences in theoretical and measurement approaches. Many of these theoretical approaches have been debated in the literature. Pearlin (1989) went as far as suggesting that disciplines should maintain their distinctive approaches to the stress process. Specifically, he argued that sociologists should avoid using medical and epidemiological models in their examination of stress. "These differences are reasonable and legitimate and they should be maintained. Sociologists should avoid immersion in the medical and epidemiological models that emphasize diagnosis and case finding. Such immersion not only fails to serve the goals of social research; it may even hinder the achievement of those goals by diverting time and resources to issues that are extraneous to social inquiry" (Pearlin, 1989, p. 253). We fundamentally disagree with this point of view and suggest that integrating such disciplinary approaches will lead to greater understanding of the stress process as similar disciplinary integration has furthered other research within human biology and other fields (Little and Haas, 1989; Rosenfield, 1992; King *et al.*, 2002; Abrams *et al.*, 2003). It can be argued that different disciplines are really just focusing on the portion of the process which best fits with their general disciplinary interests. Merging sociocultural, psychological, biomedical and evolutionary models of stress leads to a greater understanding of social, biological and cognitive components of the stress process.

"Stress etches itself into our biology and behavior, usually initiates a series of biobehavioral countering responses, and ultimately bears consequences for our social relations, ideological constructs, and evolutionary trajectories" (Goodman *et al.*, 1988, p. 170). Thus, we propose that stress be defined as a *process by which a stimulus elicits an emotional, behavioral and/or physiological response, which is conditioned by an individual's personal, biological and cultural context* (Figure 1.1). Further, the field of stress research would benefit from multiple measures across the entire process. Thus, this text guides the reader across the stress process and discusses measurement issues for the different components of the model presented in Figure 1.1. This text is intended to assist researchers in designing field-based research on stress. While these will be discussed in greater detail within each chapter, below we define and discuss in general terms the different components of the process.

Stressors

Like the rest of the stress process, stressors or stressful stimuli have been defined and categorized in several different ways. Generally speaking, they are the things that set the whole stress process in motion. Stressors are often defined as a stimulus which elicits a response; for example stressors can be defined as "external events or conditions that affect the organism" (Breznitz and Goldberger, 1982, p. 3). Wheaton notes, as others have, that the definition of a stressor is often linked to a physiological response (Wheaton, 1999). Not only is this a tautology but it is also possible that a stressor initiates a behavioral or emotional response in absence of a physiological one. He suggests that we return to an engineering concept of stress to define stressors as, "an external force acting against a resisting body" (Wheaton, 1999, p. 280). This force does not necessitate a response if an individual's resistance resources are adequate or if the force does not overload the individual's "elastic limit" (Wheaton, 1999). Stressors can be categorized along several lines, most commonly based on the temporal course or origin. For example, stressors can be categorized as acute or chronic. They may also be divided into physical, environmental or psychosocial stressors. The problem with any categorization is that it is artificial. When does a stressor become chronic for example? If the stressor occurs daily but is short-lasting each day, is that acute, chronic or somewhere in between? Equally artificial is the division based on origin. Although some physical stressors will initiate the stress process without cognitive appraisal (e.g. altitude, temperature, infection), cognitive appraisal can act to moderate the effect of the physical stressor (Mason, 1975). For example, if someone panics as they begin to have breathing difficulties at a high altitude this may exacerbate the physiological stress response. Is this then a physical or psychological stressor or both? Lazarus also divided stressors based on intensity, i.e. life events vs. daily hassles (Lazarus, 1984, 1999). In his description of stressor taxonomy of stressors, Pearlin noted,

> it should be recognized that the distinction is a construct of the researchers and does not necessarily parallel the awareness people have of their hardships and problems. People do not ordinarily sort out the various stressors that impinge on them nor do they cognitively separate eventful stressors from enduring strains [chronic stressors]. For many, the boundaries between different types of stressors become blurred as they face a mix of these stressors in the flow of their daily activities. Indeed, these boundaries can also become blurred in the eyes of the stress researcher. This is because events frequently merge into chronic strains, the strains frequently heighten the risk of stressful event. It is this tendency of events and strains to merge and blend with each other that supports the construct of stress proliferation.
>
> (Pearlin, 1999, p. 403)

Pearlin also distinguishes between primary and secondary stressors. Primary stressors are the original stressor and secondary stressors are those stressors that result from the original stressor (Pearlin, 1999). There have been numerous classifications of stressors over time; Chapter 2 discusses stressors in more detail.

Mediators and moderators

There are a variety of personal and cultural mediators and moderators that may positively or negatively affect the stress process. As with all other aspects of the stress process, different disciplines have focused on different factors. What is the difference between a mediator and a moderator? A mediator refers to a factor through which a stressor impacts the individual. A moderator somehow changes the relationship between a potential stressor and the response on an individual, "Moderating resources to control the emergence of secondary stressors, thus blocking stress proliferation" (Pearlin, 1999, p. 404). Mediators and moderators affect one's appraisal of stressors and influence the emotional, behavioral and physiological responses of individuals. A whole range of factors have been variably labeled as mediators or moderators including appraisal, personality, coping, social networks and self-concept. While these terms are often used interchangeably, the distinction is determined in analysis.* If there is an independent association between a factor and the outcome, then the factor is a mediator. If there is a statistical interaction, then it is a moderator.

The personality factors that have been most systematically examined as mediators or moderators of stress are type A & B, locus of control, John Henryism and optimism/pessimism. These personality types are largely rooted in specific cultural contexts and may not be applicable to all cultures. Personality factors are assumed to change the way that one appraises a stressor or can alter the emotional or behavioral response to stressors.

Social functioning includes social networks and social roles. There has been a great deal of work looking at the relationship between social networks and health but much of this work has been done without direct measure of physiological or emotional response. Social networks are almost always viewed as positive mediators, however, they may produce stressors in some contexts. For example, for older adults we often assume that large social networks are essential; however, older adults may find relying on others as stressful if they cannot reciprocate. We need to go beyond simply looking at the size of social networks in stress research. Dressler discusses social networks and social resources in greater detail in Chapter 2.

Historical experience of a stressor is likely to influence the appraisal of a stressor. This is rarely taken into account in stress research; however, there is evidence that repeated exposure to a particular stressor attenuates the physiological stress response. Whether this is due to physiological adaptation, a learned behavioral response or recruitment of coping mechanisms is unclear.

Mediators and moderators are also influenced by other components of the stress process including behavioral responses and an individual's health status. An individual who strengthens social networks in response to stressors has strengthened his/her adaptive capacity. On the other hand, individuals with mental or physical health problems may find themselves with a reduced social network and a loss of meaningful social roles, leaving them more vulnerable to stressors.

Coping behaviors are seen as moderators or mediating resources, "where the effects of the other components of the stress process on outcomes are channeled through the resources. Their treatment as mediators assumes that resources are not immutable but can be diminished (or replenished) by the social and economic statuses surrounding the stress process and by the ensuing stressors" (Pearlin, 1999, pp. 405–6). However, as these behaviors are often employed after a stressor is appraised as threatening, we consider them as part of the stress response.

With that said, the coping repertoire that an individual has may factor into the appraisal of a stressor, such that people who know that they have multiple coping resources may be less likely to perceive a potential stressor as a problem than those who have few coping resources.

Appraisal

Appraisal is the perception of the balance between demands and resources. With a large focus on appraisal in psychology, we would expect good tools to measure it. Unfortunately, appraisal is a nebulous concept of key importance, for appraisal determines which potential stressors result in an emotional, behavioral and/or physiological response. For example, in laboratory study of stress, math problems are commonly used as stressors. These elicit a physiological response in many individuals. However, they are unlikely to affect the same response in a person with a Ph.D. in math as in someone with only a high-school education. This is presumably due to differing appraisal of threat. Lazarus and Folkman (Lazarus and Folkman, 1984) make the distinction between primary and secondary appraisal. The primary process evaluates if a stimulus is a threat and the secondary process evaluates if the individual can meet the challenge if it is a threat. This appraisal is conditioned by cultural meaning of threat, as personal feelings of control and a variety of situational factors (e.g. previous experience with a threat or timing of event relative to the life cycle) (Lazarus and Folkman, 1984). Once a stimulus is appraised as a threat, coping behaviors can be initiated.

One way that we try to measure appraisal is to ask individuals to evaluate stressors, in other words, rate how "stressful" a particular event is. When assessing perceived levels of stress, we often confound the level of exposure to stressors, the appraisal, affective response and personal mediators. For example, the Perceived Stress Scale (Cohen *et al.*, 1983) is often used as a measure of appraisal. It asks questions such as "how often have you been upset because of something that happened unexpectedly?" This requires a respondent to sum up their exposure to unexpected things and to determine how frequently he or she gets upset by these things. In a study of long-term care residents, one participant commented, "I have learned to take things as they come" (reflecting a coping response). Several people stated, "nothing unexpected has occurred" (reflecting differential exposure to stressors) (Harper, 1998).

Appraisal of a stressor is conditioned by personal mediators, cultural and biological context. Developmental context and social variables may affect the appraisal of stimulus (Lazarus, 1984). Stressors in one culture are not necessarily appraised as stressful in others. Individuals of different age and experience will appraise events differently. Much more cross-cultural work is needed in this area. Appraisal is discussed in more detail in Chapters 2 and 3.

Stress response

The actual response to stress is multifaceted, including a behavioral, affective and physiological response which may occur individually or all at once. It is likely that they also interact with one another.

Behavioral response

When an individual is confronted with a stressor they may adopt a number of behaviors which may have a positive, negative or neutral effect on the emotional and physiological response to stressors. They may also develop behaviors to evade stressors such as early detection or avoidance. Positive behaviors, such as utilizing social networks, exercising or religious activities presumably attenuate the physiological and psychological response to stressors. On the other hand common behavioral responses to stress such as aggressive behaviors, smoking or drinking presumably accentuate the response.

Within stress literature coping behaviors are most often seen as having a positive effect on health outcomes, however, many behaviors which we employ are not particularly healthy. Many people turn to smoking, drinking or violent outbursts as a way to cope with stress. These may elevate some stress but they may also compound it. Furthermore, these negative coping behaviors may actually be the source of negative health outcomes as opposed to some physiological or affective response to a stressor. Unfortunately, researchers often statistically control for these factors rather than examining their impact as either mediators or moderators. Additionally, little work has directly linked behavioral response with physiological markers to test these assumptions. Behavioral responses to stress are further discussed in Chapter 3.

Distress or affective response

Distress refers to the emotional or affective response to stressors. There are several interview scales to assess a person's level of distress. Many of them confound the measurement of distress with appraisal and exposure to stressors (as mentioned previously with the Perceived Stress Scale [Cohen *et al.*, 1983]). Additionally, these scales are subject to recall bias. Another approach is to have participants keep diaries, in which they record their mood, behavior, and exposure to stressors. This approach, ecological momentary assessment (EMA), is not affected by recall bias and has the power of giving us an estimate of the frequency and duration of exposure to stressors, the length of an affective response, the sequence of precipitating events, and can be used to interpret physiological markers. We typically let participants define stress or anxiety for themselves in a simple question such as, "rate how stressed or anxious you feel" at specified time periods. Anxiety and stress are often used interchangeably, although they do not necessarily mean the same thing to different people. We do not know how ratings of stress on interview scales compare with data collected as diary recordings. Although one would assume that individuals who are identified as stressed in an interview will more frequently record stress or rate it more intensely, this assumption has not been well tested. One study found that data on coping collected by diary did not correlate well with data collected on interview (Schwartz *et al.*, 1999). In addition, due to participant burden, diaries are collected over a short period of time. How frequently does one have to feel distressed to result in a negative health outcome? The large majority of these studies have been cross-sectional and it is not clear if measures of distress are cross-culturally valid. Furthermore, diary recordings are challenging in illiterate and disabled populations. Affective response to stressors will be further discussed in Chapter 3.

Physiological response

The hypothalamic pituitary adrenal (HPA) axis and the sympathetic adrenal medullary system (SAMS) are activated in response to stressors. There are a variety of markers along these axes that have been measured. The two axes are believed to be reactive to different stressors but more research is needed to tease this out. Specifically, we need more studies that collect multiple markers. There are three basic approaches to research on physiological stress. In laboratory research, individuals are exposed to stressors and a physiological response is measured. Response to laboratory stressors may tell us little about real life stressors and stress responses. The ecological approach, often used in anthropology and human biology, compares groups that vary by cultural characteristics and measures levels of a physiological marker. For example, we might compare blood pressure or catecholamines between two groups that vary in level of contact with the Western market. This is a good hypothesis generating approach, however, it is prone to ecological fallacy. Exposure to stressors, personal mediators and acute physiological responses must be measured to eliminate confounding of other factors. The other approach is to collect physiological markers with diary recordings. This is a powerful approach because it links everyday experience with physiological markers. However, this approach has limitations. Because observation is time limited, the relationship between daily acute stressors and long-term chronic stress or cultural context is unclear. This approach is subject to idiosyncratic events and is limited in its ability to determine effects of chronic stress. In some cases, it is hard to interpret these data if the physiological markers and self-reported stress do not agree. The physiological response is further examined in Chapters 4–7.

Biological, environmental and cultural context

The context in which an individual interacts with stressors has the potential to alter the stress process in numerous ways. The biological context refers to the biological state of the individual. One of the most obvious aspect of biological context is developmental stage. The age of an individual will likely affect the stressors to which they are exposed and their appraisal of those stressors as well as the actual behavioral, emotional and physiological response. Not only do people habituate to stressors over time but the assessment of what is threatening depends on the developmental stage of the individual. Furthermore, we know that physiological reactivity changes over the life course (see Chapters 4, 5, 6 and 8). Men and women also respond to stressors differently, partially due to differences in appraisal and partially due to differences in physiological reactivity (see Chapters 4, 5, 6 and 8). Finally, those who are experiencing chronic stressors will respond to new stressors differently than those who are not. For example, a person who has a chronic illness is likely to have diminished psychological and physiological ability to respond to stressors compared to a healthy individual.

The physical environment can be a source of stressors. Ongoing environmental stress potentially could limit the physiological ability to respond to other stressors. For example, high altitude places a constant physiological stress (to which people are variously adapted), which may limit an individual's adaptive capacity. Few studies, however, have examined the interplay of pervasive environmental stressors and psychosocial stressors.

Cultural context influences the stressors to which people are exposed as well as the appraisal of stressors. Although there may be several universal stressors, ultimately the cultural context determines which stimuli are appraised as stressful (see Chapter 2). Furthermore, once a stressor is appraised as stressful, the behavioral and emotional response likely varies across cultures (see Chapter 3).

Health outcomes

Although stress research is ultimately interested in exploring the impact of stress on health status, biomarkers of health as opposed to disease or mortality outcomes are often used in stress research. If the particular biomarker is associated with disease or mortality risk, researchers often assume that one can logically assume that the stressor or stress response of interest also leads to increased risk. For example, the work on ambulatory blood pressure and emotion has rarely followed individuals over time to look for cardiovascular outcomes such as myocardial infarction. There has been a great deal of work looking at stress and immune response. Several authors have criticized this line of work because while research has established that stress results in predictable changes in immune function, we can't assume that it translates into more illness. This "chain of inference" often used in stress research needs to be more directly examined. We need more longitudinal studies to establish the relationships between stress and health. We do not know how much stress you need to experience to experience a negative health outcome. Further, poor health likely makes individuals more vulnerable to the stress process by affecting exposure to stressors, personal mediators, and the physiological and psychological capacity to respond to stressors. We have largely focused our attention on healthy individuals yet individuals who are ill are likely to be particularly vulnerable to stressors. While poor health is a potential outcome of the stress process, it may also increase one's risk of being exposed to stressors or may be a stressor in and of itself (Lazarus, 1984). This, in turn, may lead to an exacerbation of health or development of new conditions. For example, a diabetic may find the need for dietary restriction and repeated blood glucose testing as a chronic daily hassle, thus initiating a spiral of stress and poor health.

Which health outcomes are appropriate to study in stress research? Obviously, the answer to this question will depend on the interests of the researcher and the hypotheses which he or she wishes to test. Cassel (1976) argued that stressors can enhance disease susceptibility overall with no "etiologic specificity" (p. 111). Others have argued that some health conditions are more affected by stress than others (Lazarus, 1984). To some extent the conditions which one chooses to examine will be influenced by the stress markers chosen for the study. For example, those using ambulatory blood pressure as a marker of stress will likely look at cardiovascular disease outcomes. Furthermore, the appropriate outcome measure will be determined by the nature of stressor examined. Acute stressors are more likely to impact susceptibility to acute health conditions while chronic stressors are likely to have a greater impact on chronic disease processes. This dichotomy might break down, however, if an individual is exposed to repeated acute stressors or if exposure to a major life event alters an individual's actual or perceived exposure to chronic stressors. For example, an individual who is repeatedly exposed to daily hassles in the work place or at home may, over time, be at risk of developing chronic disease. Individuals who were subject to a traumatic life event may continuously relive such an event or have an altered perception of the world around them as threatening. Finally, there is a great need to examine culturally specific health outcomes (or cultural syndromes). For example, Kohrt *et al.* (2004) examined the epidemiology of *yadargaa*, a fatigue-related illness, in Mongolia. They found that the prevalence of *yadargaa* was associated with stressful life events. Cultural syndromes are discussed further in Chapter 2. Measurement of health outcomes is discussed in reference to specific stress markers in Chapters 2–7.

Summary

It is clear from this review that numerous disciplines have examined the stress process from a variety of perspectives. This has led to confusion over the definition of stress; however, the contrasting definitions are a result of looking at different portions of the stress process. This chapter has attempted to bring the different elements of the process together. The chapters which follow examine different parts of the process and offer practical guidelines for designing a study of stress. Chapters 2–7 [Part 2] focus on different components of the stress process, while Chapters 8–10 [Part 3] provide guidelines for overall study design. The final chapter discusses future directions for stress research.

References

Abrams, D. B., Leslie, F., Mermelstein, R., Kobus, K. and Clayton, R. R. (2003). Transdisciplinary tobacco use research. *Nicotine and Tobacco Research*, 5(Supplement 1), S5–10.

Breznitz, S. and Goldberger, L. (1982). Stress research at a cross roads. In *Handbook of Stress: Theoretical and Clinical Aspects*, ed. L. Goldberger and S. Breznitz. New York: The Free Press, pp. 3–6.

Cannon, W. B. (1914). The emergency function of the adrenal medulla in pain and major emotions. *American Journal of Physiology*, 33, 356–72.

Cannon, W. B. (1929). *Bodily Changes in Pain, Hunger, Fear, and Rage.* New York: Appleton–Century–Crofts.

Cannon, W. B. (1932). *The Wisdom of the Body.* New York: Norton.

Cassel, J. (1976). The contribution of the social environment to host resistance. *American Journal of Epidemiology*, 104(2), 107–23.

Cohen, S., Kamarck, T. and Mermelstein, R. (1983). A global measure of perceived stress. *Journal of Health, Society and Behavior*, 24(4), 385–96.

Cohen, S., Kessler, R. C. and Underwood, L. G. (1997). *Measuring Stress. A Guide for Health and Social Scientists.* Oxford: Oxford University Press.

Dohrenwend, B. and Pearlin, L. I. (1982). Report on stress and life events. In *Stress and Human Health*, ed. G. Elliot and C. Eisdorfer. New York: Springer.

Dressler, W. W. (1991). *Stress and Adaptation in the Context of Culture: Depression in a Southern Black Community.* Albany, NY: State University of New York Press.

(1995). Modelling biocultural interactions in anthropological research: an example from research on stress and cardiovascular disease. *Yearbook of Physical Anthropology*, 38, 27–56.

Dressler, W. W. and Bindon, J. R. (1997). Social status, social context, and arterial blood pressure. *American Journal of Physical Anthropology*, 102, 1343–9.

(2000). The health consequences of cultural consonance: cultural dimensions of lifestyle, social support and arterial blood pressure in an African American community. *American Anthropologist*, 102(2), 244–60.

Elliot, G. R. and Eisdorfer, C. (1982). *Stress and Human Health: Analysis and Implications of Research. A Study by the Institute of Medicine/National Academy of Sciences.* New York: Springer.

Goodman, A., Thomas, R., Swedlund, A. and Armelagos, G. (1988). Perspectives on stress in prehistoric, historical, contemporary population research. *Yearbook of Physical Anthropology*, 31, 501–6.

Harper, G. (1998). Stress and Adapations among Elders in Life Care Communities. *Anthropology*. Columbus: The Ohio State University.

Holmes, T. H. and Rahe, R. H. (1967). The social readjustment rating scale. *Journal of Psychosomatic Research*, **11**, 213–18.

Ice, G., James, G. D. and Crews, D. E. (2002). Diurnal blood pressure patterns in long-term care settings. *Blood Pressure Monitoring*, **7**(2), 105–9.

James, G. D., Jenner, D. A., Harrison, G. A. and Baker, P. T. (1985). Differences in catecholamine excretion rates, blood pressure and lifestyle among young Western Samoan men. *Human Biology*, **57**(4), 635–47.

Karlamangla, A. S., Singer, B. H., McEwen, B. S., Rowe, J. W. and Seeman, T. E. (2002). Allostatic load as a predictor of functional decline. MacArthur studies of successful aging. *Journal of Clinical Epidemiology*, **55**(7), 696–710.

King, A. C., Stokols, D., Talen, E., Brassington, G. S. and Killingsworth, R. (2002). Theoretical approaches to the promotion of physical activity: forging a transdisciplinary paradigm. *American Journal of Preventive Medicine*, **23**(Supplement 2), 15–25.

Kohrt, B. A., Hruschka, D. J., Kohrt, H. E., Panebianco, N. L. and Tsagaankhuu, G. (2004). Distribution of distress in post-socialist Mongolia: a cultural epidemiology of yadargaa. *Social Science & Medicine*, **58**(3), 471–85.

Lazarus, R. S. (1984). Puzzles in the study of daily hassles. *Journal of Behavioral Medicine*, **7**, 375–89.

(1999). *Stress and Emotion. A New Synthesis*. New York: Springer.

Lazarus, R. S. and Folkman, S. (1984). *Stress, Appraisal and Coping*. New York: Springer.

Levine, S. and Ursin, H. (1991). What is stress? In *Stress, Neurobiology and Neuroendocrinology*, ed. M. Brown, G. Koob and C. River. New York: Marcel Dekker, pp. 3–22.

Little, M. A. and Haas, J. D. (1989). Human population biology and the concept of transdisciplinarity. In *Human Population Biology*, ed. M. A. Little and J. D. Haas. Oxford: Oxford University Press, pp. 3–12.

Mason, J. W. (1974). Specificity in the organization of neuroendocrine response profiles. In *Frontiers in Neurology and Neuroscience Research*, ed. P. Seeman and G. Brown. Toronto: University of Toronto Press.

(1975). Historical view of the stress field, part II. *Journal of Human Stress*, **1**, 6–12.

McEwen, B. S. (1998a). Protective and damaging effects of stress mediators. *New England Journal of Medicine*, **338**(3), 171–9.

(1998b). Stress, adaptation, and disease. Allostasis and allostatic load. *Annals of the New York Academy of Sciences*, **840**, 33–44.

(2001). From molecules to mind. Stress, individual differences, and the social environment. *Annals of the New York Academy of Sciences*, **935**, 42–9.

(2002). *The End of Stress as We Know It*. Washington, DC: Joseph Henry Press.

McEwen, B. S. and Stellar, E. (1993). Stress and the individual. Mechanisms leading to disease. *Archives of Internal Medicine*, **153**(18), 2093–101.

Pearlin, L. I. (1983). Role strains and personal stress. In *Psychosocial Stress, Trends in Theory and Research*, ed. H. B. Kaplan. New York: Academic Press.

   (1989). The sociological study of stress. *Journal of Health and Social Behavior*, **30**(3), 241–56.

   (1999). The stress process revisited. Reflections on concepts and their interrelationships. In *Handbook of the Sociology of Mental Health*, ed. C. S. Aneshensel and J. C. Phelan. New York: Kluwer Academic/Plenum Publishers.

Rahe, R. H. (1989). Recent life change stress and psychological depression. In *Stressful Life Events*, ed. T. W. Miller. Madison: International Universities Press, Inc, pp. 5–12.

Rosenfield, P. L. (1992). The potential of transdisciplinary research for sustaining and extending linkages between the health and social sciences. *Social Science and Medicine*, **35**(11), 1343–57.

Schulkin, J. (2004). *Allostasis, Homeostasis, and the Costs of Physiological Adaptation*. New York: Cambridge University Press.

Schwartz, J. E., Neale, J., Marco, C., Shiffman, S. S. and Stone, A. A. (1999). Does trait coping exist? A momentary assessment approach to the evaluation of traits. *Journal of Personality and Social Psychology*, **77**(2), 360–9.

Seeman, T. E., McEwen, B. S., Singer, B. H., Albert, M. S. and Rowe, J. W. (1997a). Increase in urinary cortisol excretion and memory declines: MacArthur studies of successful aging. *Journal of Clinical Endocrinology and Metabolism*, **82**(8), 2458–65.

Seeman, T. E., Singer, B. H., Rowe, J. W., Horwitz, R. I. and McEwen, B. S. (1997b). Price of adaptation – allostatic load and its health consequences. MacArthur studies of successful aging. *Archives of Internal Medicine*, **157**(19), 2259–68.

Selye, H. (1936). A syndrome produced by diverse nocuous agents. *Nature*, **138**, 32.

   (1946). The general adaptation syndrome and diseases of adaptation. *Journal of Clinical Endocrinology*, **6**, 117–73.

Sterling, P. and Eyer, J. (1988). Allostasis: a new paradigm to explain arousal pathology. In *Handbook of Life Stress*, ed. S. Fisher and J. Reason. New York: Wiley, pp. 629–49.

Wheaton, B. (1999). Social Stress. In *Handbook of the Sociology of Mental Health*, ed. C. S. Aneshensel and J. C. Phelan. New York: Kluwer Academic/Plenum Publishers.

Part II
Measuring stress responses

2 *Cultural dimensions of the stress process: measurement issues in fieldwork*

WILLIAM W. DRESSLER

Introduction

The aim of this chapter is to examine measurement issues in the social and cultural study of the stress process. More specifically, the aim is to clarify a methodological orientation that can guide anthropologists and other fieldworkers interested in stress processes within specific social and cultural contexts. As such, this chapter is partly a review of how stress has been measured in many different studies, and partly an examination of the logic of measurement in anthropology and how it can be improved to understand the stress process. The chapter is intended as a supplement to the volume edited by Cohen *et al.* (1995) that reviewed issues of measurement in studies of the stress. It helps to extend that review in terms of addressing questions pertinent to the study of social and cultural dimensions of the stress process.

Culture and the stress process

In Chapter 1, Ice and James outlined the general conceptual model that guides much of research on the stress process. As they make clear, at one level it can be difficult to separate elements of the process (e.g. when is an environmental challenge an acute or a chronic stressor or when is the emotional state of the individual indicative of a stress appraisal or a coping response?). At the same time, for there to be a sensible measurement model applied to the process, conceptual distinctions need to be made.

In this chapter, I will adhere to the model outlined in Chapter 1, but with a few additions. As Ice and James suggest, an initial theoretical distinction can be made between the demanding environmental events that elicit adaptive efforts and the resources employed in adaptation. The former will be referred to as "stressors." The latter fall under Ice and James's category of "moderators and mediators," although I will make a further distinction, referring to these, following Cassel's (1976) usage, as "resistance resources" (i.e. ways of resisting the deleterious effects of stressors). These broad categories of stressors and resistance resources can be divided further. Within the category of stressors, there are those stressors that have a long-term influence on the individual and will be referred to as "chronic stressors." Then there are stressors the influence of which are relatively limited in time, and these will be referred to as "acute stressors." Within the category of resistance resources, there are resources that exist by virtue of the individual's membership in a system of social relationships. These resources will be referred to as "social resources." Then there are resources that are cognitive and emotional in nature, built up of an individual's experience of dealing with the world, as well as the experiences of dealing with the world of other persons upon which she can draw by virtue of her membership in a social support network. These will be referred to as "coping resources."

There is additionally the issue of individual appraisal. That is, irrespective of what others might think about what is occurring in the social environment, some individuals may define their experiences in terms of extreme threat and challenge. These appraisals will be referred to here as "psychological stress" and represent one level of an intermediate response to events and circumstances in the social environment. Another intermediate level is the physiologic response, in terms of changes in circulating levels of stress-related hormones, such as cortisol or epinephrine, or in terms of measures of physiologic variability, such as blood pressure reactivity.

Finally, there are the stress outcomes, or the long-term effects of these processes. These can be conceptualized as occurring at different levels. There are the long-term psychological and physical effects, such as sustained levels of depressive symptoms, or sustained high blood pressure or serum cholesterol. But anthropologists have also found other sorts of factors to be useful in studying stress cross-culturally. Some studies have examined cultural syndromes or local idioms of distress as outcomes (Oths, 1999).

Given this variety of potentially important factors, studying the stress process can be a daunting prospect. Furthermore, there are clear theoretical reasons for anticipating both nonlinear and nonadditive relationships among the stress variables and the health outcomes of

interest. Notable among these are the "buffering" model of social support, which specifies a nonadditive association between stressors and
social supports (Cassel, 1976; House and Kahn, 1985); and, the "person-
environment fit" model of stressors (French *et al.*, 1974), which specifies
a nonlinear relationship between stressors and outcomes. With other
research topics, it is possible to leave a variable or two out of consideration and, assuming little collinearity among independent variables,
you merely end up explaining a little less variance overall, but you do not
jeopardize identifying relationships of interest. In research in the stress
process, this may not be the case, in that, for example, in a given
situation the effects of stressors may be carried solely in an interaction
effect between stressors and social supports (i.e. the influence of stressors
may only be detected where social support is very low). Therefore, leaving
out a variable must be done with a clear understanding of the potential
cost.

Futhermore, <u>understanding the stress process requires understanding
how it varies under different conditions.</u> One of these conditions is that
system of shared knowledge and understanding that we call "culture."
Not long ago, there were some anthropologists arguing that the concept
of culture was nothing more than a means of reifying, essentializing, and
mystifying human differences that resulted largely from political-
economic oppression and colonial exploitation; as such, it should be
jettisoned. More recently this position has come to be seen as both
extreme and problematic, in that it removes from consideration a phenomenon that is powerful in shaping human behavior (see Brumann [1999]
for a discussion of the debate). The problems with the concept of culture
are not effectively dealt with by discarding the concept; rather, the
concept of culture must be developed, especially in terms of its utility as
an explanatory construct. What we need is a working definition of
culture, not in the usual sense of that phrase as in a heuristic definition,
but rather in the sense of a definition of culture that can do some real
work in research.

Paradoxically, the vast majority of working cultural anthropologists
proceed without having to think too much about the concept of culture.
Despite the buckets of ink spilled debating the concept, in fieldwork
there is a kind of neo-Tylorian compromise assumed that enables
ethnographers to get on with their task. Just as Tylor's (1871) famous
definition of culture subsumes everything that humans think, make,
and do, in the field, ethnographers don't think too much about what is
or is not culture, because everything is assumed to be representative
of "the culture" of the society under study. As such, culture recedes

as an explanatory construct, because everything observed is a manifestation of it.

Implicit in such a view is a cultural uniformist assumption; that is, there is a single shared culture, everybody knows it, and they get on with their lives. Under this assumption, the ethnographer's job is to decode that culture, based on observations of what individuals do and say. Although a detailed critique of such a concept of culture is beyond the scope of this chapter, there are a number of serious problems with it. These include: (a) problems with defining the substance and locus of culture, beyond euphemistic references to symbols and shared meaning; (b) an inability to deal effectively with the possibility of intracultural diversity along several dimensions; and, (c) difficulties in connecting the cultural to the individual, and hence to the biological.

Fortunately, there is an alternative to this kind of traditional concept of culture, one that can help to cope with the problems noted above, and one that can help to make the construct of culture operational. This is the perspective of cognitive anthropology (Holland and Quinn, 1987; D'Andrade, 1995; Shore, 1996). While agreement is far from complete, a basic working theory of culture in cognitive anthropology can be delineated. Drawing on Goodenough's (1996) classic definition, culture consists of that knowledge that individuals must have and share in order to adequately function in a social system. In this working theory, culture cannot be regarded as an integrated whole, but is rather made up of a set of "cultural models" that apply to various cultural domains (e.g. a cultural model of the family, a cultural model of leisure activities, a cultural model of success in life, etc.). These models (also sometimes called "schemas") are skeletal outlines of domains, consisting of the words that describe the domain and the combination of those words into prototypical descriptions of how the domain works. These models are made up of two components: one is a function of individual biography and can be thought of as a personal model; the other is a function of what the individual learns about that domain as a member of society, and hence can be thought of as a cultural model, because it is shared with other members of society (Shore, 1996, p. 49).

The notion of sharing or, as I will refer to it, "consensus," is essential in this theory of culture, as has been recognized for well over a century in the social sciences (Tylor, 1871; Berger and Luckman, 1967). Many (although probably not all) cultural models define things in the world in an essentially arbitrary way. What gives these arbitrary definitions causal force is that people agree that this is, indeed, the way things are (D'Andrade, 1984).

It is this consensus that also gives culture one of its more peculiar properties. Conventionally, culture has been regarded as a term the referent of which is an aggregate. Hence, we talk of "American culture," or "Brazilian culture." At the same time, culture must be a term the referent of which is an individual; culture cannot be located anywhere but in individuals (even if those individuals choose to use hard drives or temple walls as storage devices). Yet, culture *feels* as if it is external to us at times. And it feels that way because it actually is in one very important sense: it is in part stored in other peoples' minds. A cognitive theory of culture can locate culture in the minds of individuals, and at the same time account for the aggregate quality of culture, because that knowledge is distributed across minds. At times, certain social statuses may be the primary, if not exclusive, repository of that knowledge (e.g. priests and ritual knowledge), while at other times cultural models may simply be unevenly distributed. In either case, a theory of distributed cultural models can account for seemingly paradoxical qualities of culture.

A theory of cultural models can serve a useful role in research on the stress process. With this background, I will now turn to a selective review of some studies of stress and disease undertaken with the specific reference to understanding the stress process in social and cultural context. The emphasis will be both on what was measured and on how it was measured.

Ethnographic research on the stress process

As noted above, studies of sociocultural change or modernization and disease (e.g. Labarthe *et al.* [1973] on blood pressure, or Leighton [1959] on psychiatric disorder) serve as a foundation for more focused studies of sociocultural stressors and disease, but I will not consider these in detail here (see Dressler [1999] for a review) in favor of ethnographic studies that have tried to operationalize the model of the stress process. In order to facilitate a discussion of the logic of measurement, I will review these studies under the major categories of variables in the model of the stress process.

Chronic sociocultural stressors

One approach to the measurement of chronic social stressors is to use one's ethnographic acumen to identify conditions that are likely to be problematic for most people in a given context, and then to examine the association between the presence of such conditions and health outcomes. This describes Scotch's (1963) early study of blood pressure in South Africa. Two epidemiologic surveys were carried out in Zulu communities, one rural and one urban. Scotch examined the association of high blood pressure and various sociodemographic variables within the rural and the urban setting. For example, he found that menopausal women had higher blood pressure in the rural community, but lower blood pressure in the urban community. This he interpreted as a result of lowered social status following the loss of childbearing capacity for rural women, while for urban women, menopause freed them from the possibility of adding another mouth to feed to an impoverished household. Similarly, membership in a Christian church was associated with higher blood pressure in the rural community, and lower blood pressure in the urban community. This was interpreted as an indicator of cultural marginality in the rural area, where traditional religions predominated, but access to a community of support in the urban community, where Christian religions were normative. Scotch argued that his entire pattern of findings was consistent with a process in which adopting behaviors or assuming statuses that were not valued within a particular cultural setting was stressful for individuals, resulting in an increased risk of high blood pressure.

A similar approach was used by Rubel *et al.* (1991) in their study of *susto* in southern Mexico, and by Oths (1999) in her study of *debilidad* in the highlands of Peru. Both *susto* and *debilidad* are cultural syndromes that include combinations of chronic pain and negative mood states, and can be interpreted as culturally specific stress outcomes. Rubel *et al.* hypothesized that the risk of *susto* was a result of difficulties in carrying out culturally defined social role obligations within a community. These would include such things as a lack of participation in cooperative work groups for men, or lack of participation in church-related activities for women. They then asked individuals about their activities in these socially significant activities, and those who had fewer activities had a higher risk of *susto*. Oths, again based on ethnographic evidence, hypothesized a number of predictors of *debilidad*, including a gender imbalance within the household. She hypothesized this on the basis of ethnographic observations that, for households to function smoothly in this peasant economy, essential and gender-linked activities must be carried out. If, for example, there are few men in the household, more work falls to the women, which in turn increases their likelihood of falling ill with *debilidad* as a result of the increased set of responsibilities.

And, women in these households were found to be at a higher risk of *debilidad*.

Janes (1990) used a similar approach in his study of blood pressure among Samoan migrants to northern California. He developed a scale he called "structural stressors" based on an a priori assessment that the situations identified by the items of the scale could create substantial problems for individuals. These items included such things as chronic economic problems, difficulties with non-Samoan neighbors, and living in poor, rundown neighborhoods. This scale was not associated with blood pressure, but it was associated with a physical symptom scale, especially for women.

A comparable logic of measurement was used by Flinn and England (1997) in their studies of children's levels of circulating cortisol on the West Indian island of Dominica. These investigators were particularly interested in household composition, reasoning that stable caretaking and childrearing are essential in understanding stress levels among children. They found that children in stable households had lower levels of circulating cortisol, and what constituted a stable household was ethnographically defined. Their definition included a much wider range of household types than would be considered "stable" by American middle class standards.

A somewhat different approach to the measurement of chronic stressors can be illustrated by studies of social or status incongruence. Rather than being embedded primarily in ethnographic observation, the concept of status incongruence is derived from sociological theory on social stratification (Dressler, 2004a). The original hypothesis grew out of the observation that when there are multiple indicators of social status in a system, the same individual may occupy contradictory positions on different indicators (e.g. higher educational status coupled with lower occupational status), a state that may lead to problematic and frustrating social interaction. This body of sociological theory was then re-interpreted in terms of anthropological studies of sociocultural change and modernization. A seemingly inevitable part of the process of modernization is the introduction into local communities of new consumer goods and related behaviors that, together, can be thought of as lifestyles. In general, higher social status comes to be associated with the ability to adopt these new lifestyles. The problem, however, is that in most situations of modernization, economic growth in the form of new employment possibilities or income growth in existing occupations fails to keep pace with expanding lifestyle aspirations. There then are some people who are trying to attain and maintain the higher status lifestyle in the context of meager economic resources, a result that is likely to be frustrating and stressful (Dressler, 2004b).

This chronic social stressor has been examined in a variety of settings. In each case, two measurement issues must be addressed. The first is a lifestyle inventory. Typically this is based on general ethnographic observation. The quality of a lifestyle inventory is thus dependent on the sensitivity of the ethnographer to local definitions of status. For example, in my own work on the West Indian island of St. Lucia (Dressler, 1982), I developed a brief inventory of items such as owning a home, having plumbing, owning a radio, owning a stereo, and similar basic consumer goods. Additionally, I included the item "glass case," which is a simple glass-doored cabinet in which can be displayed painted plates and glasses. The glass case is a potent symbol of a kind of middle class propriety in St. Lucia, and possession of it clearly indicates that an individual is trying to communicate a sense of his or her place in the world. Then, I assessed economic resources with an occupational status ranking. A specific form of status incongruence, that I referred to descriptively if somewhat infelicitously as "lifestyle incongruity," was operationalized as having a higher than median lifestyle with a lower than median occupational status. Higher blood pressure was associated with lifestyle incongruity. Other investigators have replicated this finding in a variety of settings and in relation to a variety of outcome variables, but it is important to emphasize that the specific operational indicators of lifestyle and economic resources must be chosen in terms of ethnographic realities (McGarvey and Schendel, 1986; Janes, 1990; Dressler, 1991a,b; Dressler *et al.*, 1992; Chin-Hong and McGarvey, 1996; Bindon *et al.*, 1997; McGarvey, 1999; McDade, 2001; 2002).

Other studies using this general concept are helpful in emphasizing this point. For example, in his study of Samoan migrants to California, Janes (1990) argued that migrants experienced different kinds of status incongruence. One involved attempts to achieve social status within Samoan-defined terms, which included attaining the title of *matai* (or, roughly, "chief") and a leadership role within the community. The ability to carry out these social roles depends in part on having economic resources, so Janes examined the discrepancy between this leadership scale and income, and he referred to this as "intra-community social inconsistency." The other form of status incongruence involved attempts to achieve status in American-defined terms, which Janes measured using a fairly typical sociological approach as a discrepancy between occupational status and educational level. He found both forms of social inconsistency to be associated with higher blood pressure for men, but not for women.

More recently, McDade (2001) has examined lifestyle incongruity and immune status among Samoan adolescents. This research presented an interesting methodological issue in that status incongruence was assessed for individuals (i.e. adolescents) for whom status was, at least in part, derived from others (their parents). He found that adolescents who lived in a household with a higher status lifestyle, as assessed by an inventory of material goods, showed evidence of greater infectious disease challenge if their parents had lower occupational and educational status. He also developed a slightly different measure of status incongruence. In this measure, he assessed first the degree of exposure of an adolescent to information outside the local community (as measured by exposure to media, travel, and friendships with non-Samoans). He also determined if the household in which the adolescent lived had a member who had the higher status of *matai*. If there was a discrepancy between degree of exposure outside the community and the social status of the household on the basis of the chiefly status, the adolescent was more likely to show evidence of infectious disease challenge (McDade, 2002).

Another approach to measuring chronic social stressors is to ask individuals to report their experiences along a number of dimensions. For example, in research on stress, depression, and blood pressure in an African American community in the rural, southern USA (Dressler 1991a), I adapted a set of social role stressor scales originally developed by Pearlin and colleagues (Pearlin *et al.*, 1981) for use in a community study of depression carried out in the urban North. Individual stressor scales were designed to elicit from each respondent his or her perceptions of chronic difficulties in major social roles, including that of economic provider (financial stressors), husband or wife (marital stressors), being a parent (parenting stressors), the world of work (job stressors, unemployment stressors), and other relationships (stressors associated with being single). I started with the items in each scale (there were 8–12 items per scale) as they had been phrased by Pearlin in the original study; then, working with a group of three research assistants who had grown up in the study community, I adapted the phrases and meaning of the items to the local vernacular. This resulted in some items that would be unlikely to appear in standard scales, but which spoke to the everyday life of members of the community. For example, one item in the economic stressor scale read: "How often do you feel you have more month left than money?" This of course was a phrase used commonly in the community to describe a strain on resources. All of these scales then demonstrated acceptable internal consistency reliability

(Cronbach's alpha > .70), and all were associated with depressive symptoms (but, interestingly, not with blood pressure, see Dressler, [1991b]).

A similar approach was used by Oths and her associates in their study of psychosocial stressors and low birth weight in a community in the southern USA (Oths *et al.*, 2001). Oths started with a brief ethnographic study, carrying out open-ended interviews with pregnant women in a clinic serving a low income neighborhood. She found that women complained a great deal about the nature of work during pregnancy. The majority of women were working in low income, entry-level jobs in convenience stores or fast-food outlets, or were working in light industry such as chicken processing plants. In these jobs they had little control over their work, especially in terms of being able to take breaks or to make or receive telephone calls. Oths was struck by the parallels between these kinds of demands on the job and the demands described by Karasek (1996) in his demand–control model of occupational stress. This model of occupational stress identifies stress as resulting from the imbalance between demands on the job and the degree of control that an individual has over those demands. The difference was that Karasek had developed his model primarily in terms of men in white-collar occupations, and Oths was developing a model for blue-collar women. As such, she had to translate the items from terms applicable to the stereotypical junior executive to terms applicable to the convenience store clerk. This was done by modeling the language of the items closely on the way that key informants had spoken of their jobs. Oths *et al.* (2001) found that psychosocial job stressors were prospectively associated with having a lower birthweight baby, especially for African American women.

Janes (1990) also used this approach in his study of Samoan migrants. He developed a "family stressors" scale that consisted of "... events in the household or extended family that command extraordinary attention" (p. 124). These included such things as a death in the extended family, arguments with family or church members, and unexpected expenses connected to the extended family. Family stressors were associated with higher blood pressure, primarily for women.

Palinkas *et al.* (1993a,b) explored chronic stressors in European American and Alaskan Native communities in the aftermath of the Exxon Valdez disaster. They developed a scale of chronic stressors to assess the degree to which residents of communities that varied in their exposure to the disaster perceived a decline in the quality of social relationships as a function of the accident and its cleanup. They also developed a scale of the direct economic impact of the accident.

They found that European American community residents were more likely to report depressive symptoms if they also suffered economically as a direct result of the disaster; Alaskan Natives, on the other hand, reported more depressive symptoms if they also perceived a decline in the quality of social relationships in their communities.

Acute sociocultural stressors

The term "acute sociocultural stressors" refers to events that are often sudden and unexpected and, as such, can pose considerable adaptive demands. The most well-known approach to the measurement of these stressors, usually referred to as "stressful life events," was developed by Holmes and Rahe (1967). They generated a long inventory (more than 40 such events) ranging from common and seemingly mundane events (such as getting a parking ticket) to events likely to generate severe crises in a family, such as the death of a spouse. They then asked a "normative" sample to rate the degree to which the occurrence of each event was likely to require some kind of readjustment on the part of the individual experiencing the event. The weighted list of events became the "Social Readjustment Rating Scale." In research, an investigator could simply determine what events had occurred to an individual, and then sum the weights to arrive at the adaptive demand.

The literature on stressful life events is vast, with very little of it generated in cross-cultural research. It is a literature, however, in which anthropological concerns have loomed large, even if anthropologists have contributed little to the discussion. The main issue has been: what is more important, the occurrence of an event that is generally regarded as stressful in a society (which is the logic of the original scaling), or an individual's subjective evaluation of the occurrence of the event (which harks back to an understanding of stress exclusively in terms of individual appraisal)? Generally speaking, there is little evidence that subjective evaluations of life events make much difference; the vast majority of the explanatory variance is captured by the occurrence of a relatively few dramatic and highly culturally-salient events that disrupt the social fabric of an individual's life within a particular social setting (Tausig, 1982).

For example, I examined stressful life events in research on depression in an African American community. Based on ethnographic evidence, I argued that there were likely to be different kinds of events with special meaning for different segments of the community. Given the strong emphasis on work in terms of the social definition of a good person in the community, and especially the way in which this moral definition of personhood loomed large in the older (age > 40) age group, I hypothesized that the event of unemployment by itself would be a significant correlate of depressive symptoms for that age group. I then hypothesized that noneconomic life events (getting married, birth of a child, death of a family member, starting a new job) would be more important for other age groups. The correlations were consistent with these hypotheses (Dressler, 1991a).

Janes (1990) also examined stressful life events in his study of Samoan migrants. He gathered data on major events involving significant loss, such as a child leaving home, loss of a job, or an accident to or significant illness of a family member. The occurrence of these events were associated with increased number of health complaints for both men and women.

The importance of examining stressful life events that are meaningful within a local context can be further illustrated by McDade's (2003) research on immune function among adolescents in Samoa. McDade developed an inventory of life events, one of which was the number of extended family obligations the household had had in the year prior to the interview. He found that the association of life events and immune function depended on the economic status of the household. Where economic status was high, more events were associated with better immune function; where economic status was low, more events were associated with worse immune function.

The researcher who has probably done most to bring an anthropological sensibility to the measurement of stressful life events is the British medical sociologist George Brown (Brown, 1974; Brown and Harris, 1978). Brown took a very dim view of approaches that used the subjective assessment of the stressfulness of an event by the research subject, arguing that, in practice, it would be impossible to distinguish between the perceived stressfulness of the event and efforts by the individual to cope with the occurrence of the event. For example, it has been shown that individuals under extreme stress may try to cope with the stressful event by minimizing the conscious impact using classic defense mechanisms such as suppression or denial. Their self-report of the impact of the event could then be precisely the opposite of its true impact. Brown argued instead that the assessment of the impact of an event needed to be based on the overall context in which it occurred. For example, loss of a job may be but a minor inconvenience for someone with the skills to become quickly re-employed, while for someone with few marketable skills, it could be catastrophic. Brown developed a system for rating the impact of an event by having a panel of raters come to an agreement on the likely impact, given their phenomenological under-standing of the person's life, excluding any information on the indivi-dual's own subjective assessment. The question posed was: how would a reasonable person in this social situation be expected to respond to this event? Life events assessed in this way have been shown to be potent predictors of the onset of major depression (Brown, 1974; Brown and Harris, 1978).

Social resources

As with stressors, two approaches to the measurement of social resources have been employed. On the one hand, researchers have identified social arrangements in which individuals are likely to have access to help and support in times of felt need. Then, the degree to which individuals are integrated into these social arrangements is assessed. Generally, these are referred to as measures of "social integration." On the other hand, individual's perceptions of the availability of help within their social network in relation to specific problematic events have been assessed. This perception or belief in the availability of help and support has been referred to as "social support" (Cohen, 1988).

A number of studies have employed measures of social integration. In many respects this lends itself to traditional anthropological concerns with kinship and social structure. For example, on the West Indian island of St. Lucia, it is rare for individuals to marry early in life. Instead, it is common for persons to have more-or-less long-term relationships with two or more persons before they marry, relationships that usually result in common children. As household composition shifts, children stay with their mothers, but ideally paternity is never denied nor forsaken, and good men are expected to support their own children to some extent, even though they may reside in other households. What this results in, in effect, is a complex network of households linked by common children, between which resources of various kinds flow. Individuals who have children in other households, and who are further linked by virtue of a network of adult siblings, thus are part of a network of mutual support, and have lower blood pressure (Dressler, 1982).

Janes (1990) used a similar logic in his study of Samoan migrant blood pressure. Kinship is a complicated subject in Samoan society, and extended kin and the ritual social obligations entailed by extended kinship can be very demanding. At the same time, there is a core set of kin, usually consisting of adult siblings, but sometimes including other kin and close friends, that can reliably be identified as a significant social support network. Janes counted the number of these relationships for subjects in his study, and found the larger core social networks were associated with lower blood pressure for both men and women (Janes, 1990).

With respect to social support, a number of studies have employed an approach in which respondents are asked hypothetical questions about whom they could or would turn to in specific situations requiring help or assistance. This is a fairly common way of looking at ego-centered social networks, and when that is the focus, respondents are asked to name actual people in their network. A number of studies of social support have taken a somewhat broader view, however, by naming particular kinds of social relationships. So, for example, in a study in Mexico, I asked respondents to whom they could turn in response to specific problems, and I included relatives, friends, and neighbors as alternatives; I also included the category *compadre* or *comadre*, given the continuing importance of the system of ritual kinship known as *compadrazgo* in Mexico. Results indicated that men (but not women) who perceived their *compadres* as a source of support in a variety of situations had lower blood pressure; women who perceived greater family support had lower blood pressure. These results are consistent with ethnographic analyses of Mexican village social structure (Dressler, Mata *et al.*, 1986).

I employed this same approach in a study of social support and blood pressure in an African American community (Dressler, 1991a, 1991b). Extended kin are traditionally very important sources of social support in the black community; however, since the impact of the economic changes brought about by the Civil Rights Movement have come to be more widely felt, the importance of the extended family has changed. Many younger black people share more problems with friends and co-workers than they do with their extended kin, especially as new opportunities in education and employment have opened to them. Therefore, I hypothesized that there would be two dimensions of social support that were important, and that these two dimensions in turn would have different effects for younger and older members of the community. I found that, for younger persons, higher non-kin support moderated the impact of status incongruence on depressive symptoms and blood pressure. For older persons, higher kin support moderated the impact of status incongruence on depressive symptoms and blood pressure.

These latter results illustrate the importance of embedding an understanding of social support within the larger context of social structure, and they illustrate the importance of examining nonadditive effects as well as linear and additive effects. In this specific ethnographic case, neither kin nor non-kin social support had an association with blood pressure, but in interaction with status incongruence, the two factors were important within particular age groups.

McDade (2001) examined the modifying effects of social relationships on immune function in his research in Samoa. Rather than ask about support per se, he asked adolescents to rate their satisfaction with their social relationships. When this was examined in interaction with lifestyle incongruity, adolescents who rated their social relationships as more satisfying *and* who had higher lifestyle incongruity had evidence of compromised immune function.

Overall, anthropological studies of social integration, social support, and health outcomes point to the importance of embedding the measurement of these factors in an understanding of local pattern of social structure and social organization (Dressler, 1994).

Coping resources

As noted above, coping resources are those personal characteristics of adaptability, resilience, and resistance that individuals use to deal with the occurrence of problems. These coping resources have also been thought of as "styles," or characteristic patterns, of approaching problems.* Coping can be divided into instrumental coping, referring to attempts to directly alter a stressful event or circumstance, and emotion-focused coping, referring to attempts to manage one's emotional reaction to an event or circumstance. Relatively little work has been done cross-culturally on this topic, while there is a vast literature in psychology. It is useful to note two examples from the psychological literature, simply to bracket the range of possible approaches to the issue. On the one hand, there is Kobasa's (1979) concept of the "hardy personality type," which is a personality configuration including traits such as internal locus of control, high self-efficacy, and high self-esteem. In broad outline, the hardy personality sounds like someone who is self-confident and ready to take charge in any given situation and who, not surprisingly, is resistant to the effects of stressors. On the other hand, there is the "Ways of Coping Inventory," a list of over 100 possible ways that one might respond to a given situation, such as "try to forget about it," or "try to get more information about it" (Lazarus and Folkman, 1984). The effectiveness of ways of coping will then depend on the nature of the specific stressor. For example, it would be problematic to try to change something that cannot be changed (e.g. the death of a relative); instead, a more emotion-focused strategy would be optimal in that given situation (see Chapter 3 for additional discussion of coping instruments).

These two approaches to the study of coping are useful in illustrating the way the concept is viewed in the literature. In one sense, it is a broad and relatively enduring characteristic of persons; in another sense, it is a set of discrete strategies that can be selected in any given situation.

As I noted, there is very little research on coping in the anthropological literature, and most of it tends to view coping as a style or characteristic pattern for individuals. In my research in St. Lucia, I developed a sentence-completion test of coping styles. The sentence stems consisted of incomplete statements such as "When I have a problem I ..." or, "When someone tells me I cannot do it I ... ." The respondent is then asked to complete the sentence in his or her own words. Statements such as "When I have a problem ... I try to solve it" can be coded as direct-action coping (similar to the more general concept of instrumental coping), while statements such as "When I have a problem ... I just forget about it" can be coded as a passive or emotion-focused coping. Again, coding these responses must be sensitive to local idioms for talking about life. In St. Lucia, I coded the response "prayer" as direct-action coping, because the term referred to a conscious effort to develop a strategy to deal with a problem that involved requesting divine advice, not as a withdrawal into a passive, external locus of control kind of inaction. I found that individuals who reported a more direct-action style of coping also had lower blood pressure (Dressler, 1984).

I used the same sentence-completion test in research in Brazil and in the African American community in southern USA. In Brazil, I found that Afro-Brazilians who expressed a direct-action coping style had blood pressures as low or lower than European Brazilians, presumably because they were better able to deal with issues of racism and economic marginality (Dressler, Santos and Viteri, 1986). Results in the African American community were quite different. In that study I coded the sentence-completion test more explicitly as instrumental versus emotion-focused coping. I hypothesized that men in the African American community in the rural South would be more comfortable with an emotion-focused style of coping, because of historic patterns of racial violence against assertive black men. I further hypothesized that women could effectively employ a more instrumental style of coping, because historic patterns of sex role differentiation led their coping efforts to be focused more within the household and family. The results indicated that an instrumental coping style did buffer the effects of economic stressors on psychological and physical symptoms for women, while an emotion-focused coping style buffered the effects of economic stressors on psychological and physical symptoms for men. These results again point to the need to understand well the ethnographic context in proposing, analyzing, and interpreting stress hypotheses (Dressler, 1985).

McDade (2003) has used a similar kind of variable in his research on immune function among Samoan adolescents. He examined characteristic patterns of coping, divided into "self-reliant" versus "socio-reliant" categories. To assess these, he asked adolescents if, in response to a series of problems, they would tend to deal with it themselves, or they would seek out the help of someone else. He found a buffering effect of coping on stressful life events in relation to immune function, such that life events had a smaller effect on immune function for those employing a self-reliant strategy.

Psychological stress

It should be apparent now that in the scheme presented here, the generalized subjective concomitants of stressful events and circumstances, including an amalgam of thoughts and emotions such as feeling out of control, angry, anxious, and depressed, is considered to be a variable of possible interest, but it is not considered to be "the" essential variable that must be measured. In large part, this is because, as noted above, it may be difficult to differentiate in practice such a response from efforts to cope. Nevertheless, it can be very interesting to examine a variable called psychological stress. There is a long tradition of using symptom checklists, such as the venerable Cornell Medical Index (CMI) as a measure of stress (e.g. Ness, 1977; McGarvey and Schendel, 1986). Usually the assumption is made that the symptoms listed in the CMI are either culturally neutral or such basic reports of symptoms that translation per se is sufficient to use it in different settings. Such an assumption may be warranted in some settings, while in others cultural models of the working of the body and mind may be so different as to render the questions about symptoms problematic.

Another example of transporting a psychological stress scale to another setting is my use of Cohen's Perceived Stress Scale (Cohen *et al.*, 1983). This is a 10-item scale that assesses perceived stress in terms of feelings of a lack of control in most situations and feelings of anger and frustration at that lack of control. When translated into Brazilian

Portuguese, this scale appears to function as intended, as evidenced by significant associations with depressive symptoms and with cultural consonance (a construct to be discussed in detail below) in the expected directions (Dressler *et al.*, 2002). More careful examination of the meaning of the scale is still necessary, however.

I experimented with the use of a stressful life event inventory as a measure of perceived psychological stress in my research in St. Lucia. Using an inventory of life events, I asked respondents what the impact on them would be if the event were to happen, reasoning that individuals who saw themselves as more vulnerable to life events were experiencing greater distress. There were no correlations with blood pressure; however, the pattern of the rating of the life events was instructive. Respondents with high blood pressure rated a cluster of culturally salient life events as having more impact than respondents with normal blood pressures (Dressler, 1984).

One of the most interesting approaches to the measurement of psychological stress is Ice's work among elderly members of the Luo ethnic group in Kenya (Ice and Yogo, 2005). Due to the impact of HIV/ AIDS, many elderly people function as caregivers, both for ill family members and for children who have lost their parents to the disease. She elicited terms from key informants that could be considered to be reasonable translations of the term "stress" in the local language. She then asked people to list all of the psychological and somatic concomitants of this condition. Combining those terms, she developed a culturally specific psychological stress scale. This scale is clearly applicable only to this language and cultural group, but it can be taken as a very sensitive indicator of psychological stress for intracultural comparisons. More work of this type needs to be done.

Anthropological studies of the stress process: some implications

As I noted at the outset, this has been a selective rather than extensive review of studies of the stress process in cross-cultural perspective, the aim of which has been to illustrate how anthropologists have gone about the task. At this point, I will draw out some of the implications of these studies with respect to studying stress in the field. First, while few of the studies reviewed discuss the concept of culture that provides a foundation for the research, the theory of cultural models discussed earlier is a useful way to interpret many of the results. For example, the studies of chronic social stressors that identify conditions that are likely to prove problematic for most people in a community, and then relate those conditions to outcomes, assume implicitly that there is a shared understanding of that cultural domain, and what are the preferred states of variables associated with that domain. Oths's (1999) study of *debilidad* is instructive in this regard. Identifying a sex-ratio imbalance in the household as a potential stressor depends on an understanding of cultural models of family life and the division of labor in the peasant household. Furthermore, it presumes that this is a shared and meaningful model of everyday life for the participants themselves. Assuming this meaningfulness then enables Oths to hypothesize that deviations from this prototypical model of family life and the division of labor will be experienced as problematic for individuals, which in turn will be related to health outcomes. Underlying the whole research strategy is an assumption of these shared understandings.

Similarly, in my use of stressful life events as a measure of acute stressors in the African American community (Dressler, 1991a), I argued that different events are differentially meaningful for different persons, because the cultural models in which those events are embedded are differentially distributed in the community. For older African Americans, in open-ended interviews, the moral dimension of work was emphasized repeatedly. Of course work provided economic resources, but beyond that, work signified the contribution of the person (and especially men) to the well-being of their families and the well-being of their community. To be a person was to be a working person. Clearly, unemployment could and would be problematic for younger persons as well, but as an event it would not carry the same valence as social transitions involving marriage, starting a family, or beginning and ending one's education. Therefore, I argued that unemployment and noneconomic life events would have different effects for different age groups, based on an understanding of intracultural diversity in the distribution of cultural models of work and social life.

From this review, it is clear that anthropological studies of the stress process are most effective when they concentrate on first identifying the collective meaning and understanding of cultural domains that are particularly salient in specific contexts, and then determine how different individual experience within those domains may affect health.

A specific understanding of cultural models of, for example, family life or lifestyle is necessary, but not sufficient, for studying stress in the field. Virtually all of the studies reviewed above depended also on a broader understanding of the ethnographic context to fully account for the results. Janes's (1990) study of Samoan migrants to northern California is instructive in this regard. Samoan social life is structured by shared models of extended kinship. A superficial understanding of Samoan society could lead to a treatment of the kinship network as an undifferentiated whole. Janes found, however, based on his intensive ethnographic study of the community, that there were two distinct dimensions in the kinship structure. On the one hand, the extended kin system could be very demanding in terms of both the time and economic resources of individuals and households. Involvement in this broader social system, when not coupled with sufficient economic resources, became a measure of a stressful intracommunity social inconsistency. On the other hand, Janes identified core members of the kin group, especially adult siblings, as a source of help and support, and the number of those in an individual's social network became a measure of social support. Again, a detailed understanding of the broader ethnographic context is essential in this research.

Finally, as I have noted elsewhere, successfully testing these hypotheses often requires the proper specification of complex multivariate statistical models (Dressler, 1995). For example, Bindon *et al.* (1997) examined the lifestyle incongruity hypothesis in American Samoa. In carrying out the research, one of the ethnographers on the project, based on her observations, suggested that the effects of this form of status incongruence might differ for men and women. Lifestyle is fundamentally a way of displaying to the world an individual's ability to accumulate the visible indicators of middle class success, a cultural model of everyday life that quickly permeates local communities in the process of sociocultural change and modernization. But often this particular form of success is of greater importance for men rather than women. In many societies, men are the public representatives of the household and hence carry in interaction with the outside world the responsibility of status display. But if a status claim is asserted and not supported by status in other dimensions (e.g. occupational status, educational status, *matai* status), such a claim can be seen as fraudulent and rejected by others in social interaction. For women in Samoa, status revolves around their responsibilities as wives and mothers, in their ability to maintain a family and household in a way collectively regarded as good. So, Bindon *et al.* (1997) hypothesized that lifestyle incongruity would be more strongly associated with blood pressure for men rather than women, a hypothesis that was confirmed (and that replicated Janes's results from the migrant Samoan community). Detailed discussions of multivariate statistical models would be beyond the scope of this chapter; however, it should be noted that specifying this statistical model required, in essence, estimating a three-way interaction effect in which one of the two-way interactions was modeled as a discrepancy term and the three-way interaction term was a cross-product of the discrepancy term with gender. Other studies (e.g., Janes, 1990; Dressler, 1991a; McDade, 2003) also involve the proper specification of quite complex statistical effects. The main point, however, is that these specifications are ethnographically defined.

New directions in research: a theory of cultural consonance

In this concluding section of this chapter, I will turn to some recent work that promises to open new avenues of investigation for biocultural research. This research is more explicitly linked to the culture theory outlined above, and it incorporates new innovations in research methods in cultural anthropology. Furthermore, it is somewhat less specifically tied to the model of the stress process outlined above, although the general concept of stress is employed within it.

As noted above, the notion that some set of circumstances or events might be stressful or supportive for individuals depends on the assumption that these events or circumstances have meaning for them, and that this meaning is shared. While cultural sharing or consensus has long been assumed in cultural anthropology, in the past 20 years efforts have been made to specifically test the degree to which individuals do indeed share an understanding of a particular cultural domain. Romney *et al.* (1986) introduced the cultural consensus model, which accomplishes the essential task of quantifying consensus. Working from the pattern of agreement among key informants, the cultural consensus model determines precisely the degree of sharing in a domain. The degree of consensus in a domain enables the analyst to infer within certain confidence limits that these informants are, or are not, operating from a shared cultural model. Additionally, the cultural consensus model can operationalize the degree to which individuals in a sample share in the overall consensus. This is the concept of "cultural competence," which is the correlation between an individual's understanding of the domain and the consensus understanding of the domain. Finally, the cultural consensus model enables the analyst to estimate the "culturally best" set of responses within a particular domain ("best" in the sense that these are more likely to reproduce more individuals' responses). The responses are estimated by giving higher weight to the informants who have higher cultural competence (or, in other words, who can replicate more closely the group-level responses (Romney *et al.*, 1986). This estimate is not a simple average, but rather it takes into account the way in which meaning is distributed among the informants. In other words, it is a function not of what any individual knows, but rather of how that knowledge is distributed.

One criticism of a cognitive view of culture is that it tends to deal only with the way things are thought to be, and not with action. Put differently, people don't just know or think things, people *do and believe* things, and assessing the degree to which individuals conform in their behaviors and their personal beliefs to cultural prototypes for those behaviors and beliefs is an important question (Crossley, 2001). This may be especially important in studies of health outcomes. Therefore, I have proposed an additional theoretical construct and measure that can link the cultural to the individual. This is the concept of "cultural consonance," and it is defined as the degree to which an individual approximates in his or her own behavior or belief the collective prototypes for that belief or behavior in a particular domain, as these are encoded in a cultural model.

It is hypothesized that low cultural consonance is a stressful experience for individuals. If a cultural domain has been identified that is central to the social life of a community, then an individual who has low cultural consonance in that domain is, by definition, at the margins of the valued patterns of behavior in the community. This can be a frustrating and stressful experience for the individual, especially assuming that an understanding of the preferred patterns of belief and behavior are widely shared.

The concept of cultural consonance was developed in two studies that examined both arterial blood pressure and depressive symptoms. In Brazil, Dressler *et al.* (1997, 1998) examined cultural consonance in two cultural domains: lifestyle and social support. They argued that lifestyle (consisting of the accumulation of material goods and the adoption of related behaviors) could be viewed as a way in which claims to social status in mass society are expressed in mundane social interaction. Lifestyle is, in essence, a performative dimension of socioeconomic status. Persons who are seen to be closer in their approximation to cultural models of a successful lifestyle are accorded higher social status, which in turn is associated with more satisfying mundane social interactions. Social support, on the other hand, is an essential social resource for dealing with the inevitable crises, large and small, of everyday life. However, social support systems are not constructed solely out of voluntary relationships; rather, there are categories of persons to whom it is more (or less) appropriate to appeal for assistance in particular situations (Dressler, 1994). Therefore, it is likely that cultural models of social support guide individuals' choices in this regard, and closer approximation to these models may in turn lead to more satisfying social interactions.

In carrying out cultural consensus analysis, Dressler *et al.* (1997, 1998, 1999) worked with inventories of lifestyle items, and with combinations of everyday problems and potential social supporters, that had been developed using traditional ethnographic methods. They found consensus on lifestyle items deemed indicative of being a success in life, and on the pattern of resort to types of persons in relation to specific problems. Both of these consensus models were tested in a sample of key informants ($n = 20$) designed specifically to detect intracultural diversity (if such diversity was important). They then collected data on these same items in a survey sample ($n = 304$) on which they also collected blood pressures and other data. They found that those individuals who more closely approximated the cultural consensus model in their own behaviors in these two domains — that is, persons who were higher in cultural consonance — also had lower blood pressures (after controlling for a variety of known and suspected correlates of blood pressure). Low cultural consonance has also been found to be associated with an increased risk of depressive symptoms, heavy drinking, and total caloric intake (Dressler *et al.*, 2002, 2004).

The model was also tested in a study in an African American community in southern USA (Dressler and Bindon, 2000) using essentially the same methods outlined above. Virtually the same results were obtained, although there was a statistical interaction effect between the two measures of cultural consonance in the US study, while these were simple main effects in the Brazil study.

In more recent work, the model of cultural consonance has been expanded in two ways in research in Brazil. First, a much more complete set of methods for examining the cultural domains of interest have been employed. Second, a wider variety of domains themselves have been examined (Dressler *et al.*, 2006). With respect to domains, the new research, in addition to lifestyle and social support, includes the domains of family life, national character, and food. In order to understand these domains better, a more complete set of methods of cultural domain analysis was employed (see Weller and Romney, 1988; de Munck and Sobo, 1998; and Ross, 2004 for detailed discussions of these techniques). For example, in each domain, we first asked key informants ($n = 43$) to free list elements of the domain. In the domain of lifestyle, this involved asking them to list all of the things necessary for living a good life.

In the domain of national character, we asked them to list all of the characteristics that make Brazilians Brazilian. Depending on the domain, this task generated anywhere from 60–100 terms. These were then reduced to 25–30 core terms, based on eliminating redundancy and using those terms employed by a greater proportion of the informants (although care must be taken not to discard terms that may be infrequently used, but nevertheless interesting).

We then turned to a set of tasks to explore the meaning of the elements of the domains. For example, we asked a set of informants ($n = 20$) to carry out unconstrained pile sorts of the elements of the domains. This involves transferring each term to an index card, and then asking informants to group terms that are similar, making as many piles as are necessary. The results of a pile sort can be converted to a similarity or proximity matrix, and the results displayed graphically. Also, interviewing informants after the task can reveal the dimensions of meaning they used to think about the domain. For example, in the domain of lifestyle, informants seemed to use an evaluative dimension of "importance," or those things that are truly important to having a good life, versus those things that might be nice to have, but are not truly important. Similarly, in the domain of food, informants used a dimension of "health" to group food items, as well as dimensions of "convenience" and "prestige." In the domain of family life, an evaluative dimension of good versus bad families was salient, while for national character, there was a great deal of contention about what were or were not truly Brazilian characteristics.

An additional task was used to examine these dimensions more closely. Based on the pile sorts, another sample of informants ($n = 32$) was asked to rank the elements along various dimensions, but most importantly an evaluative dimension. These rankings could then be correlated with interpoint distances from the pile sort task to confirm the dimensions of meaning. For each of the cultural domains, an evaluative dimension was most important, although other dimensions could be important as well (notably in the domain of food).

It was at this point that we turned to the testing of cultural consensus in each of the domains, using a larger ($n = 66$) and more representative sample. For lifestyle, family life, and national character, informants were asked to rate the importance of each element for defining the domain. For food, informants ranked the elements along five dimensions. In no case were they asked to rate them in terms of their own personal models, but rather in terms of general community standards (e.g. "Rate the importance in this community of these lifestyle items for having a good life"). The emphasis was on what informants understood to be shared community standards, and not on what they idiosyncratically thought. These ratings were then examined using cultural consensus analysis. There was consensus for all domains, although it ranged from very low (national character) to extremely high (lifestyle, the health dimension of food). The cultural answer keys for the domains then represented the best estimate of a culturally appropriate rating of the item. All of the data management and data analysis routines necessary for these kinds of data are available in the program ANTHROPAC (Borgatti, 1993).

At this point, we turned to the development of scales that could assess the degree to which individuals approximate the pattern of belief and behavior described in the consensus model in their own lives. These scales would then be used in an epidemiologic survey in which data on cardio-vascular and mental health were also obtained. In the domains of lifestyle, social support, and food, this posed little problem. That is, we could ask what lifestyle items an individual possessed; if they would seek social support in the pattern described by the cultural model; and, what foods they had eaten over the past two weeks. In each case, reported patterns of behavior could be matched to the consensus ratings of the importance of the items.

In the domains of family life and national character, however, there are complications in measurement at the individual level. Direct questions about the concepts important in the domains of the family and national character seem more difficult (e.g. it is difficult to ask "Do you really love your family?"). Therefore, we used a time-honored tradition in the social sciences: we asked respondents about their perceptions, phrased in terms of statements with which they could agree or disagree. For example, the concept of *compreensão* or understanding within the family, was extremely important, and in the survey sample we posed the following statement: "In my family we understand one another completely." Respondents were then asked if they agreed or disagreed with that statement. At least one statement was created for each of the concepts in each domain. So, for the domains of lifestyle, social support, and food, we have scales of how well reported behaviors match the consensus models of the domain. For the domains of family life and national character, we have scales of individual perceptions of how their family or of how their own behaviors do or do not correspond to the cultural models of the domains.

All of the measures of cultural consonance achieve acceptable levels of internal consistency reliability (Cronbach's alphas $> .70$) and have been found to be associated either with blood pressure or with depressive symptoms. A detailed examination of the results can be found in other papers. For our purposes here, it is sufficient to note that cultural consonance appears to be a potent correlate of both physical and mental health outcomes (Dressler, 2005; Dressler, Borges *et al.*, 2005; Dressler, Balieiro *et al.*, 2005).

One of the advantages of the method outlined in the operationalization of cultural consonance is that it is explicitly linked to a specific culture theory (see Figure 2.1 for a schematic of the measurement model). There is an unbroken line that can be drawn from the concept of culture to the measurement of cultural consonance at the individual level. This is essential if the cultural dimensions of the stress process are to be understood. The ethnographic studies of the stress process reviewed earlier do this as well, but making the case that, for example, a particular measure of social support is culturally valid in a particular setting depends very much on the ethnographic skills of the investigator, as well as on his or her rhetorical skills. By that I mean that the reader of the research report must be convinced that the measurement "makes sense" within a particular ethnographic setting; this usually depends on carefully setting the ethnographic context and presenting a coherent description of that setting, such that there is a convincing logical coherence to all of the results.

This logical coherence then creates the sense that the ethnographer understands the environment of shared meaning and understanding in which the study took place, and that the measurements applied order individuals along various continua in valid and reliable ways. The method outlined in the study of cultural consonance removes some of the

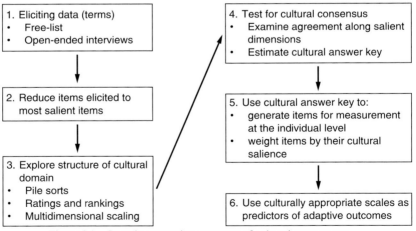

Figure 2.1. Steps in generating measures of cultural consonance.

rhetorical weight from this argument. Instead, in this method, public and replicable techniques of data elicitation and analysis are applied to first examine the shape and structure of the environment of meaning in a particular setting. Once that environment of meaning is described, then the distribution of individuals around the culturally defined central tendency can be assessed, and these measurements can be evaluated with the usual tools of psychometric analysis. The essential point, however, is that the system of shared meanings has been examined first.

There are few studies employing this method outside of my own research group at this point in time; however, two other investigators have used related approaches. Decker's study (Decker *et al.*, 2003) of cortisol among Dominican men fits nicely with this approach. He identified several dimensions by which men in Dominica evaluate each other, including helpfulness, trustworthiness, and influence. He then had men in the village rate each other along these dimensions. Men who were agreed to fit the prototypical cultural model of being a man in that society had lower average cortisol levels and reported fewer depressive symptoms.

Handwerker (1999) used a similar approach in the study of depressive symptoms among working class women in the urban northeastern USA. A set of stressors was generated from free listing and open-ended interviewing, including such things as being demeaned, made to feel inferior, blocked in achievement efforts, or otherwise led to feel bad about oneself, and the shared understanding of these events or circumstances was confirmed using consensus analysis. Then the frequency of the stressors was assessed and found to be associated with depressive symptoms.

Finally, this method can be made to work in the opposite direction. That is, given an existing scale, the data-elicitation and analysis techniques can be used to determine if the items of the scale are culturally meaningful within the community being studied. Gannotti and Handwerker (2002) used this approach in research on disabilities in Puerto Rico. They found that a widely used scale of disability could be adapted with little modification in Puerto Rico. The method of cultural domain analysis enabled these researchers to evaluate the scale much more precisely.

Summary

Studies of the stress process have proven to be a very productive avenue of research for anthropologists and other social scientists interested in cultural dimensions of human biology. In this chapter, two general approaches have been reviewed, with special reference to research methods employed. The first general approach derives its inspiration from the syntheses of the model of the stress process that appeared in the 1970s and 1980s. In this model, broad classes of variables were specified that included the stressors that demand adjustive response by individuals, as well as the social and psychological resources that individuals use to respond. In studies employing this approach, it is clear that much more specification must take place within a given ethnographic context. What counts as a chronic stressor, or what counts as a social support, can vary considerably depending on how these elements are culturally defined. It was argued as well that the meaning of the phrase "culturally defined" is best understood in terms of a theory in which culture is conceptualized as a set of vertically and horizontally linked models of discrete cultural domains.

Identifying stressors and resistance resources within any given setting is only part of the task. Culture not only defines what a stressor or a social support is in a given setting, it also can modify the expected associations among variables. For example, McDade (2001) anticipated that social integration would moderate the effects of lifestyle incongruity on immune function in the way predicted by the buffering model of social support. He found instead that higher social integration exacerbated (rather than ameliorated) effects of lifestyle incongruity. There are of course a variety of factors that could alter this direction of effects, not the least of which is the focus on immune function in adolescents as opposed to blood pressure in adults. But, it could be that the environment of meaning in which these processes occur alter their effects. As McDade notes: "Although social relationships in Samoa have their undeniable benefits, they may not always be synonymous with social *support*, at least as conceptualized in previous studies of lifestyle incongruity" (McDade, 2001, p. 1359, italics in the original). Again, making the case depends on the researcher's overall knowledge of the ethnographic context, and being able to convey the logical coherence of the findings with respect to that context.

This is not necessarily problematic; indeed, it is nothing more nor less than anthropologists have always done. But, as in any scientific endeavor, the more empirical links that can be filled in, the better. This is the importance of the explicit theoretical orientation of cultural models, along with the data elicitation and analysis techniques of cognitive anthropology. The environment of cultural meaning can be explicitly and systematically explored using these techniques. In my discussion of this approach above, I examined the approach primarily in terms of studies of cultural consonance. It should be emphasized here that a theory of cultural consonance can be separated from the application of the method used to study it. I happened to be interested in this dimension of cultural systems, and it turns out that this dimension of cultural systems has implications for human biology as well. The general method could easily be used in the study of elements of the stress process. Take, for example, stressful life events. It would be straightforward to elicit from informants events that cause major and minor problems in everyday life, and to explore the structure of such a domain via pile sorts and ranking tasks. Then, informants could be asked to rank the life events in terms of the degree to which they would cause problems in everyday life, and cultural consensus could be used to determine if there is an agreed upon model. If so, the investigator would then have a culturally valid inventory of stressful life events ready for use in that society, which in turn could be compared to those used in other settings (see Dressler *et al.* [1991] for a discussion of comparing culturally appropriate measures across societies). My main point, however, is that these methods of cultural domain analysis are general and flexible in their application. Continuing refinement of anthropological methods in research on the stress process will result in a better understanding of the general issue of human adaptability.

Acknowledgment

Preparation of this chapter was supported in part by research grant BCS-0090193 from the National Science Foundation.

References

Berger, P. L. and Luckman, T. (1967). *The Social Construction of Reality*. New York: Doubleday.

Bindon, J. R., Knight, A., Dressler, W. W. and Crews, D. E. (1997). Social context and psychosocial influences on blood pressure among American Samoans. *American Journal of Physical Anthropology*, **103**, 7–18.

Borgatti, S. P. (1993). *Anthropac 4.05*. Harvard, MA: Analytic Technologies.

Brown, G. W. (1974). Meaning and the measurement of life events. In *Stressful Life Events: Their Nature and Effects*, ed. B. S. Dohrenwend and B. P. Dohrenwend. New York: Wiley, pp. 217–44.

Brown, G. W. and Harris, T. (1978). *Social Origins of Depression*. New York: Free Press.

Brumann, C. (1999). Writing for culture: Why a successful concept should not be discarded. *Current Anthropology*, **40**(Supplement 1), s1−27.

Cassel, J. C. (1976). The contribution of the social environment to host resistance. *American Journal of Epidemiology*, **104**, 107−23.

Chin-Hong, P. V. and McGarvey, S. T. (1996). Lifestyle incongruity and adult blood pressure in Western Samoa. *Psychosomatic Medicine*, **58**, 130−7.

Cohen, S. (1988). Psychosocial models of the role of social support in the etiology of physical disease. *Health Psychology*, **7**, 269−97.

Cohen, S., Karmack, T. and Mermelstein, R. (1983). A global measure of perceived stress. *Journal of Health and Social Behavior*, **24**, 385−96.

Cohen, S., Kessler, R. C. and Gordon, L. U. (1995). *Measuring Stress: A Guide for Social Scientists*. New York: Oxford University Press.

Crossley, N. (2001). The phenomenological habitus and its construction. *Theory and Society*, **30**, 81−120.

D'Andrade, R. G. (1984). Cultural meaning systems. In *Culture Theory: Essays on Mind, Self and Emotion*, ed. R. A Schweder and R. A. Levine. Cambridge: Cambridge University Press, pp. 88−119.

(1995). *The Development of Cognitive Anthropology*, Cambridge: Cambridge University Press.

Decker, S., Flinn, M., England, B. G. and Worthman, C. M. (2003). Cultural congruity and the cortisol stress response among Dominican men. In *Social and Cultural Lives of Immune Systems*, ed. J. M. Wilce, Jr. London and New York: Routledge, pp. 147−69.

de Munck, V. C. and Sobo, E. J. (1998). *Using Methods in the Field: A Practical Introduction and Casebook*. Walnut Creek, CA: Altamira Press.

Dressler, W. W. (1982). *Hypertension and Culture Change: Acculturation and Disease in West Indies*. South Salem, NY: Redgrave Publishing.

(1984). Hypertension and perceived stress: a St. Lucian example. *Ethos*, **12**, 265−83.

(1985). The social and cultural context of coping: action, gender, and symptoms in a Southern black community. *Social Science and Medicine*, **21**, 499−506.

(1991a). *Stress and Adaptation in the Context of Culture: Depression in a Southern Black Community*. Albany, NY: State University of New York Press.

(1991b). Social support, lifestyle incongruity, and arterial blood pressure in a Southern black community. *Psychosomatic Medicine*, **53**, 608−20.

(1994). Cross-cultural differences and social influences in social support and cardiovascular disease. In *Social Support and Cardiovascular Disease*, ed. S. A. Shumaker and S. M. Czajkowski. New York: Plenum Publishing, pp. 167−92.

(1995). Modelling biocultural interactions in anthropological research: an example from research on stress and cardiovascular disease. *Yearbook of Physical Anthropology*, **38**, 27−56.

(1999). Modernization, stress and blood pressure: new directions in research. *Human Biology*, **71**, 583−605.

(2004a). Social or status incongruence. In *The Encyclopedia of Health and Behavior*, ed. N. B. Anderson. Thousand Oaks, CA: Sage Publications, pp. 764–7.

(2004b). Culture, stress and cardiovascular disease. In *The Encyclopedia of Medical Anthropology*, ed. C. R. Ember and M. Ember. New York: Kluwer, pp. 328–34.

(2005). What's *cultural* about bio*cultural* research. *Ethos*, **33**, 20–45.

Dressler, W. W., Balieiro, M. C., Ribeiro, R. P. and Santos, J. E. D. (2005). Cultural consonance and arterial blood pressure in urban Brazil. *Social Science and Medicine*, **61**, 527–40.

(2006). Cultural consonance and psychological distress. *Culture, Medicine and Psychiatry*, **30**, in press.

Dressler, W. W., Balieiro, M. C. and Santos, J. E. D. (1997). The cultural construction of social support in Brazil: associations with health outcomes. *Culture, Medicine and Psychiatry*, **21**, 303–35.

(1998). Culture, socioeconomic status and physical and mental health in Brazil. *Medical Anthropology Quarterly*, **12**, 424–46.

(1999). Culture, skin color, and arterial blood pressure in Brazil. *American Journal of Human Biology*, **11**, 49–59.

(2002). Cultural consonance and psychological distress. *Paidéia: Cadernos de Psicologia e Educação*, **12**, 5–18.

Dressler, W. W. and Bindon, J. R. (2000). The health consequences of cultural consonance: cultural dimensions of lifestyle, social support and arterial blood pressure in an African American community. *American Anthropologist*, **102**, 244–60.

Dressler, W. W., Borges, C. D., Balieiro, M. C. and Santos, J. E. D. (2005). Measuring cultural consonance. *Field Methods*, **17**, 331–55.

Dressler, W. W., Evans, P. and Gray, D. J. P. (1992). Status incongruence and serum cholesterol in an English general practice. *Social Science and Medicine*, **34**, 757–62.

Dressler, W. W., Mata, A., Chavez, A. and Viteri, F. E. (1986). Social support and arterial blood pressure in a central Mexican community. *Psychosomatic Medicine*, **48**, 338–50.

Dressler, W. W., Viteri, F. E., Chavez, A., Grell, G. A. C. and Santos, J. E. D. (1991). Comparative research in social epidemiology: measurement issues. *Ethnicity and Disease*, **1**, 379–93.

Dressler, W. W., Santos, J. E. D. and Viteri, F. E. (1986). Blood pressure, ethnicity, and psychosocial resources. *Psychomatic Medicine*, **48**, 509–19.

Flinn, M. and England, B. G. (1997). Social economics of childhood glucocorticoid stress response and health. *American Journal of Physical Anthropology*, **102**, 33–53.

French, J. R. P., Rogers, W. and Cobb, S. (1974). Adjustment as person-environment fit. In *Coping and Adaptation*, ed. G. V. Coelho, D. A. Hamburg and J. E. Adams. New York: Basic Books, pp. 316–33.

Gannotti, M. and Handwerker, W. P. (2002). Puerto Rican understandings of child disability: methods for the cultural validation of standardized measures of child health. *Social Science and Medicine*, **55**, 11−23.

Goodenough, W. (1996). Culture. In *The Encyclopedia of Cultural Anthropology*, ed. D. Levinson and M. Ember. New York: Henry Holt, pp. 291−9.

Handwerker, W. P. (1999). Cultural diversity, stress, and depression: working women in the Americas. *Journal of Women's Health & Gender-Based Medicine*, **8**, 1303−11.

Holland, D. and Quinn, N. (1987). *Cultural Models in Language and Thought*. Cambridge: Cambridge University Press.

Holmes, T. H. and Rahe, R. H. (1967). The social readjustment rating scale. *Journal of Psychosomatic Research*, **11**, 213−18.

House, J. S. and Kahn, R. L. (1985). Measures and concepts of social support. In *Social Support and Health*, ed. S. Cohen and S. L. Syme. Orlando, FL: Academic Press, pp. 88−108.

Ice, G. H. and Yogo, J. (2005). Measuring the emotional toll of the HIV/AIDS crisis on Luo grandparents: development and psychometric properties of the Luo Perceived Stress Scale (LPSS). *Field Methods*, **17**, 394−411.

Janes, C. R. (1990). *Migration, Social Change and Health: A Samoan Community in Urban California*. Stanford, CA: Stanford University Press.

Karasek, R. (1996). Job strain and the prevalence and outcome of coronary artery disease. *Circulation*, **94**, 1140−41.

Kobasa, S. C. (1979). Stressful life events, personality, and health: an inquiry into hardiness. *Journal of Personality and Social Psychology*, **37**, 1−11.

Labarthe, D., Reed, D., Brody, J. and Stallones, R. (1973). Health effects of modernization on Palau. *American Journal of Epidemiology*, **98**, 161−75.

Lazarus, R. S. and Folkman, S. (1984). *Stress, Appraisal, and Coping*. New York: Springer Publishing.

Leighton, A. H. (1959). *My Name is Legion*. New York: Basic Books.

McDade, T. W. (2001). Lifestyle incongruity, social integration, and immune function among Samoan adolescents. *Social Science and Medicine*, **53**, 1351−62.

(2002). Status incongruity in Samoan youth: a biocultural analysis of culture change, stress and immune function. *Medical Anthropology Quarterly*, **16**, 123−50.

(2003). Life event stress and immune function in Samoan adolescents. In *Social and Cultural Lives of Immune Systems*, ed. J. M. Wilce, Jr. London and New York: Routledge, pp. 170−88.

McGarvey, S. T. (1999). Modernization, psychosocial factors, insulin and cardiovascular health. In *Hormones, Health and Behavior*, ed. C. Panter-Brick and C. M. Worthman. Cambridge: Cambridge University Press, pp. 244−80.

McGarvey, S. T. and Schendel, D. E. (1986). Blood pressure of Samoans. In *The Changing Samoans: Behavior and Health in Transition*, ed. P. T. Baker, J. M. Hanna and T. S. Baker. New York: Oxford University Press, pp. 350−93.

Ness, R. C. (1977). Modernization and illness in a Newfoundland community. *Medical Anthropology*, **1**, 25−53.

Oths, K. S. (1999). *"Debilidad:"* A reproductive-related illness of the Peruvian Andes. *Medical Anthropology Quarterly*, **13**, 286–315.

Oths, K. S., Dunn, L. K. and Palmer, N. S. (2001). A prospective study of psychosocial job strain and birth outcomes. *Epidemiology*, **12**, 744–6.

Palinkas, L. A., Downs, M. A., Petterson, J. S. and Russell, J. (1993a). Social, cultural, and psychological impacts of the Exxon Valdez oil spill. *Human Organization*, **52**, 1–13.

Palinkas, L. A., Petterson, J. S., Russell, J. and Downs, M. A. (1993b). Community patterns of psychiatric disorders after the Exxon Valdez oil spill. *American Journal of Psychiatry*, **150**, 1517–23.

Pearlin, L. I., Menaghan, E. G., Lieberman, M. A. and Mullan, J. T. (1981). The stress process. *Journal of Health and Social Behavior*, **22**, 337–56.

Romney, A. K., Weller, S. C. and Batchelder, W. H. (1986). Culture as consensus: a theory of culture and informant accuracy. *American Anthropologist*, **88**, 313–38.

Ross, N. (2004). *Culture & Cognition*. Thousand Oaks, CA: Sage.

Rubel, A. J., O'Nell, C. W. and Collado-Ardon, R. (1991). *Susto: A Folk Illness*. Los Angeles and Berkeley: University of California Press.

Scotch, N. A. (1963). Sociocultural factors in the epidemiology of Zulu hypertension. *American Journal of Public Health*, **53**, 1205–13.

Shore, B. (1996). *Culture in Mind*. New York: Oxford University Press.

Tausig, M. (1982). Measuring life events. *Journal of Health and Social Behavior*, **23**, 52–64.

Tylor, E. B. (1871). *Primitive Society*. Boston, MA: Estes & Lauriat.

Weller, S. C. and Romney, A. K. (1988). *Systematic Data Collection, Vol. 10, Qualitative Research Methods Series*. Newbury Park, CA: Sage.

3 *Measuring emotional and behavioral response*

GILLIAN H. ICE

Introduction

This chapter aims to review measures of the emotional and behavioral responses to stressors. As reviewed in Chapter 1, there are multiple cognitive responses to a stressor starting with the appraisal of a stressor. After a person appraises a stressor or evaluates it as a threat, s/he will experience a stress response, often including an emotional and behavioral response. For example, an individual is exposed to the stressor of caring for a relative with dementia, s/he may feel burdened, anxious and sad. This same person may engage in various coping behaviors such as smoking or seeking solace at a place of worship. The way that an individual responds to a stressor (from the appraisal through the stress response) depends on various factors (e.g. personality, coping resources) which may mediate or moderate the way in which an individual appraises and then responds to a stressor. For example, this same caregiver, may have a large supportive family which minimizes the appraisal of threat or the emotional response in comparison to an individual without such support. The reader should refer to Chapters 1 and 2 for the definition of the concepts of mediators, moderators, appraisal and behavioral and emotional responses as this chapter will focus on the measurement approaches and will briefly review published measures of such concepts and different measurement approaches. One should note, as discussed in Chapters 1 and 2, that often these cognitive processes are lumped together in a single measure of "perceived stress."

Interview methods

The most common way for researchers to obtain data on the behavioral and emotional responses to stress is by interview. There are numerous stress and affect scales that have been well validated on US and European populations. Many of them have been only validated with a select group of participants (largely young adults, often college students and often white and middle class). These have typically not been validated outside developed countries. Researchers should be cautious about choosing a scale "off the shelf" and assuming that the questionnaire is valid outside the tested population. It is not a simple matter of translation. Many words or phrases may have no meaning in non-Western contexts. Even within Western populations, there may be cohort effects such that certain concepts are foreign or inappropriate for some age groups. So while many of the scales presented below have been well validated within certain populations, you cannot use them in other contexts without testing to establish validity and reliability (described later in the chapter). Published scales do provide an excellent starting point and would likely be less time consuming to validate than creating a new scale. For a more comprehensive review of published scales, readers should refer to Cohen *et al*. (1997), particularly the chapters by Monroe and Kelley (1997) and Stone (1997).

Published scales

Appraisal

Appraisal is defined as the perception of the balance between demands and resources. As discussed in detail in Chapter 1, appraisal is critical to the stress process; however, few good scales exist to measure appraisal (Monroe and Kelley, 1997). A number of studies use what Monroe and Kelley (1997) refer to as ad hoc measures. These are single items (or a limited number of times) designed to assess the response to a specific situation. With this approach, individuals might be asked to assess on a Likert scale how stressful an event is or if they are able to cope with an event. This approach is most typically used with laboratory studies or with diaries (Monroe and Kelley, 1997). An alternative approach is to ask an individual to recount a stressful event and describe their appraisal using these ad hoc measures. For example, Marlowe (1998) asked headache sufferers to keep a diary of stressful events. For each event participants were asked to assess the "extent to which the event mattered to them" (primary appraisal) and "the extent to which they believed they could change the event" (secondary appraisal) (Marlowe, 1998, p. 1110). Morris and Long (2002) asked clerical workers to recount their most salient work stressor within the previous month. Primary appraisal was assessed by asking them to rate the extent to which the event was upsetting, important, a threat to work competence and a threat to self-esteem. Secondary appraisal was assessed by asking them the extent to which they had control over the stressor. Chang (1998) asked psychology students to appraise an exam by answering the following questions: "how important was the event," "how much control did you feel you had over the outcome," "how effectively did you feel you were able to prepare for the event," "how much stress did it evoke," "how threatening was the event," and "how challenging was the event." In a more extensive approach, Dewe (1991) asked insurance company employees to describe their most stressful event in the previous month. Following an approach by Folkman *et al.* (1986), they were then asked eight questions to assess primary appraisal (e.g. feeling you would not achieve an important goal; feeling you would lose the respect of someone important to you; feeling threatened; appearing to be unsupportive, etc.). Four questions were used to assess secondary appraisal (one that you could change or do something about; one that must be accepted or just got used to; one where you had to hold yourself back from doing what you wanted to). There are many more examples of this approach in the literature. This approach is limiting as it is often context specific and is rarely psychometrically tested. Furthermore, psychometric properties are expected to be poor with a single or limited number of items. Finally, it should be noted that several of these items appear to confuse appraisal with coping or emotional response.

There are three multi-item scales which have been referred to as appraisal measures: the Stress Appraisal Measure (SAM), the Perceived Stress Scale (PSS) and the Perceived Stress Questionnaire (PSQ). SAM was designed following the Lazarus/Folkman model of primary and secondary appraisal (Peacock and Wong, 1990). Peacock and Wong (1990) identified three dimensions of primary appraisal: threat, challenge and centrality. Threat refers to the potential for harm, challenge refers to the anticipation of gain and centrality refers to the perceived importance of the event. Secondary appraisal focused on perceptions of control – controllable by self, controllable by others and uncontrollable by anyone. The scale was designed with subscales to tap into these six dimensions as well as overall perceived stressfulness. With the exception of the uncontrollable subscale, the subscales were found to have good internal consistency. All subscales were confirmed by factor analysis and were found to have good criterion validity among undergraduate psychology students. Although this scale has been used by other researchers in different contexts, it has not been well tested in other populations. In a sample of adolescents, only three factors of appraisal were identified using SAM (Rowley *et al.*, 2005). The Stress Appraisal Measure can be found at http://www.twu.ca/cpsy/Documents/wong/SAM.pdf.

The PSS has been described as a "global measure of stress" (Cohen *et al.*, 1983). It was designed to measure an individual's overall appraisal of the stressfulness of their lives. The scale was designed to "tap the degree to which respondents found their lives unpredictable, uncontrollable and overloading." The authors state that the scale is appropriate for those with a junior high school education or higher. The scale was initially validated with two college samples and one smoking cessation group but since the initial publication the scale has been widely validated and used. Ten and four-item versions of the scale have also been published (Cohen and Williamson, 1988). The PSS is probably the most commonly used measure of perceived stress and has been shown to be predictive of a number of health outcomes (Cohen and Lichtenstein, 1990; Cohen *et al.*, 1993; Cohen, 1995; Martin *et al.*, 1995; Cohen *et al.*, 1999). The scale has been translated into Spanish (European and Mexican), Thai and Chinese; however, on his website, Dr. Cohen states that the translated versions have not been tested by his lab nor is he aware of psychometric properties. Although the PSS has been described as a measure of appraisal, it includes items that can be labeled as affective response to stressors or coping. The PSS was found to have two factors (perceived distress and coping) in two samples of psychiatric patients (Hewitt *et al.*, 1992; Martin *et al.*, 1995). Multiple versions of the scale and references can be found on Dr. Cohen's webpage http://www.psy.cmu.edu/~scohen/.

The PSQ has also been described as a measure of appraisal; however, the scale items include affective responses. The authors of the scale state that it was designed "for clinical psychosomatic research" and items were drawn from experienced clinicians of patients with "chronic relapsing diseases regarding their life situations that actually trigger symptoms" (Levenstein *et al.*, 1993, p. 20). The scale was validated with clinic samples of English and Italian speakers and college students. It was found to be correlated with trait anxiety, self-reported stress, somatic symptoms and the PSS. Seven factors were identified (harassment, overload, irritability, lack of joy, fatigue, worries and tension). The scale was subsequently examined with a larger, more diverse sample and found to have four factors (worries, tension, joy, demands) (Fliege *et al.*, 2005). Even by the names of the factors it is clear that the scale is primarily measuring affective response, therefore this scale should probably not be used as an appraisal measure but might be considered for a measure of affective response. The PSQ was translated into Spanish and validated in clinic and healthcare samples in Spain (Sanz-Carrillo *et al.*, 2002). The PSQ is published in the original publication (Levenstein *et al.*, 1993).

Affective response

Psychologists have long discussed the theory and structure of mood. The current debate over the proper classification is beyond the scope of this book (interested readers should consult Plutchik and Kellerman [1989]; Lazarus [1991]; Clark [1992]; Stone [1997]; Watson and Vaidya [2003]; however, the classification of mood does have some impact on its measurement. The most popular classification of mood is based on the circumplex model. In this model, based on factor analysis, there are two independent factors, positive and negative affect. Furthermore, moods are arranged in a circle across two major dimensions, high−low activation (arousal or engagement) and pleasantness−unpleasantness (valence or evaluation). Emotions which are arranged closely to one another are assumed to be similar to one another; those that are at 90° from one another are independent and those which are at 180° are inversely related. Therefore, a person's negative affect does not directly affect their positive affect. However, the model has been criticized partially based on the labeling of factors. Specifically, what makes a mood negative: high activation or low pleasantness (Lazarus, 1984; Larsen and Diener, 1992)? Furthermore, people have noted that by grouping emotions, unique features of individual emotions are lost (Lazarus, 1991; Larsen and Diener, 1992; Stone, 1997). Finally, the inverse relationship between positive and negative affect decreases as the time span shortens (Dienner and Emmons, 1985). Although there are some differences, the overall structure of mood appears to hold across cultures (Shaver *et al.*, 1992).

Another common distinction between moods is state and trait effects. State refers to the variability of mood at any given time, whereas a trait refers to a person's emotional disposition over a longer period of time. Typically in stress studies, moods assessed by interview are traits whereas those collected by diaries are states.

Few scales have been developed which specifically examine the affective response to stressors. More typically, investigators use validated scales to measure affect or single-item measures of moods of interest ranked on a Likert scale (e.g. anxiety, anger, happiness, sadness). Adjective checklists are most commonly used (Plutchik, 1989; Stone, 1997). There are numerous scales from which to choose (Stone, 1997;

Watson and Vaidya, 2003) but more commonly used measures are the Positive-Affect and Negative-Affect Schedule (PANAS) (Watson *et al.*, 1988) and the Profile of Mood States (POMS) (McNair *et al.*, 1971). The PANAS asks individuals to rate the extent to which they felt a series of emotions on a five-point Likert scale. There are ten positive moods (interested, excited, strong, enthusiastic, proud, alert, inspired, determined, attentive, active) and ten negative moods (distressed, upset, guilty, scared, hostile, irritable, ashamed, nervous, jittery, afraid). This measure can be used as a state or trait measure depending on the time-course instructions given to the participants. The PANAS has been well tested and has sound psychometric properties (Watson and Vaidya, 2003). The POMS contains 65 mood adjectives and participants are instructed to rate the extent to which they've felt each emotion on a five-point scale. There are six subscales (tension—anxiety, anger—hostility, confusion—bewilderment and vigor—activity). It can be used with various time instructions. The POMS has been well tested and has good psychometric properties (Watson and Vaidya, 2003).

Behavioral response

Generally, the measurement of the behavioral response to stressors is referred to as coping. The concept of coping has taken on different meanings but Lazarus and Folkman (1984a) note that they all share a central theme: "the struggle with external and internal demands, conflicts, and distressing emotions" (p. 283). They formally define the psychological concept of coping as, "the process of managing demands (external or internal) that are appraised as taxing or exceeding the resources of the person" (p. 283). Although often conceived as positive behaviors, coping behaviors need not be positive nor effective. For example, many people smoke to "cope" with stressful events, however, smoking is not positive and probably not particularly effective at managing stressors. There are a handful of coping measures based on different theoretical approaches to coping (Cohen, 1991; Schwarzer and Schwarzer, 1996). The most commonly used general assessments of coping are the Ways of Coping Questionnaire (WCQ) (Folkman and Lazarus, 1980) and the COPE Scale. The WCQ, originally known as the Ways of Coping Checklist, has undergone revision and a significant amount of psychometric testing. It consists of 50 items, which are rated on a four-point Likert scale, divided into eight subscales based on factor analysis (confrontive coping, distancing, self-controlling, seeking social support, accepting responsibility, escape-avoidance, planful problem solving, positive reappraisal). The WCQ scale and manual can be purchased at http://www.mindgarden.com/products/wayss.htm (an older, 67-item version is published in Lazarus and Folkman [1984b]). The COPE scale was developed using two theoretical models, the Lazarus model of stress and the "behavioral self-regulation model" (Carver *et al.*, 1989). The scale consists of 13 theoretically derived subscales (active coping, planning, suppression of competing activities, restraint coping, seeking social support — instrumental or emotional, positive reinterpretation and growth, acceptance, turning to religion, focus on and venting of emotions, denial, behavioral disengagement, mental disengagement and alcohol–drug disengagement). The entire scale is published in the original paper (Carver *et al.*, 1989).

Creating a culturally specific scale

While stress research has long been conducted in Western societies, focus on psychosocial stress among non-Western, non-industrialized societies is relatively new. As a result, most scales designed to measure affective or behavioral responses to stressors have been developed for and tested on Western populations. It is possible to use or adapt scales for use in other cultures, but it may be appropriate in many cases to create a culturally specific scale. Even though scales developed for use within a specific cultural context are not as useful for cross-cultural comparisons, they may be better at measuring the impact of local stressors on the stress response and health outcomes. Thus, the goal of the research project must be considered before deciding on a behavioral or emotional measure of stress. Anthropologists and psychologists have noted over the years that culture has an impact on mental health. Studies on the etiology and expression of schizophrenia (Edgerton, 1966, 1971, 1980; Katz *et al.*, 1988; Edgerton and Cohen, 1994), depression (Kaaya *et al.*, 2002) and post-traumatic stress syndrome (Breslau, 2004; Derluyn *et al.*, 2004; Pham *et al.*, 2004) have made clear that culture shapes the factors which lead to mental illness as well as the expression of mental illness. For example, symptoms of deviant or inappropriate behavior common to schizophrenia are defined by cultural concepts of normality and rules of appropriate behavior (Jenkins and Karno, 1992). Prevalence of mental illness, symptom clusters and treatment also differ across cultures (Edgerton, 1966, 1971, 1980; Katz *et al.*, 1988; Edgerton and Cohen, 1994).

While most of the research on culture and mental health has focused on mental illness, several authors have noted that culture impacts the way in which emotions are normally expressed (Myers, 1979; Jenkins and Karno, 1992). Jenkins and Karno (1992) note, "cultures differentially construct a universe of discourse on emotion, or ethos ... . Emotions that are culturally salient (e.g. sadness as opposed to anger) provide models that may shape how individuals might or should feel in a given situation" (p. 17). These conclusions are useful for researchers as they adopt scales for measuring psychosocial stress in non-Western contexts.

In many languages the words "stress" or "stressed" are not used but the absence of the word does not imply that the stress process does not exist. It does mean, however, that perceived stress and mental health scales created for western populations may be inappropriate for use in non-Western populations (Patel *et al.*, 1999; Patel, 2001; Pike and Young, 2002). Common feelings identified in Western scales may be inappropriate in other cultural contexts (Pike and Young, 2002). The Perceived Stress Questionnaire (Levenstein *et al.*, 1993) asks "how often do you feel under pressure from deadlines?" or "feel frustrated?" and the Perceived Stress Scale (Cohen *et al.*, 1983) asks "how often have you felt nervous and stressed?" or "how often have you felt that things were going your way?" These expressions may have no meaning in some cultures. Therefore, the use of published scales can result in an under- or over-estimation of distress, and often result in missing data due to non-response (Pike and Young, 2002). Patel and colleagues (Patel *et al.*, 1999; Patel, 2001) and Pike and Young (2002) have argued that to adequately assess emotions and mental health in different cultures, it is necessary to identify local "idioms of distress" as the basis of a culturally specific scale.

Although there are many reasons why investigators may choose to create a new scale, researchers must be aware of the tremendous time and financial commitment necessary for such an effort. Creating a new scale or questionnaire is not simply a matter of generating questions. Researchers must take several steps to ensure reliability and validity of a new scale. Scale development is a multiple-step process which requires numerous tests and revisions (Spector, 1992). In many circumstances, the adaptation of pre-existing scales may be preferable, but these are subject to similar pitfalls if the structure of a complex construct such as perceived stress varies across cultures. Regardless of the path an investigator chooses – adaptation of existing scales or creation of new scales – she or he must be prepared for a long and challenging process. Scale development is not a project to take on for a dissertation!

Scale development

There are numerous texts on scale development and psychometrics to which the reader should refer (for example, Carmines and Zeller, 1979; Converse and Presser, 1986; Weller and Romney, 1988; Spector, 1992; Fowler, 1993, 1995; DeVellis, 2003; Netemeyer *et al.*, 2003). This section will provide a general review of the process. The scale construction process consists of five steps: 1) defining the concept, 2) designing the scale, 3) pilot testing, 4) administration and item analysis, 5) validation and norming (Spector, 1992). The process of scale development must start with the definition of the construct to be measured. Without a clear concept a scale cannot be designed. In measuring concepts related to stress, the investigator must be clear on what part of the process s/he chooses to measure.

To begin the process of scale design, the response choices must be determined. Agreement and frequency are common choices in stress scales (Spector, 1992). Agreement asks participants to rate the extent to which they agree with a statement (somewhat agree, agree, somewhat disagree, disagree). Frequency evaluation asks participants to rate how frequently something has occurred (never, rarely, sometimes, often). Responses are typically ordered and assigned an ordinal which can be summed across items. Items can be ordered to be unipolar (all going in one direction) or bipolar (going in two directions with a zero point) (Spector, 1992).

After you decide on a scoring system, you must generate a list of items for a scale (words or phrases). These can come from ethnographic data, focus groups or from other scales. Ethnographic data sources include published resources, key informant interviews or the systematic data collection methods described by Dressler in Chapter 2 (for additional information see Weller and Romney, 1988; de Munck and Sobo, 1998). Focus group methodology is beyond the scope of this text; however, it should be noted that it is not simply a matter of chatting with a group of people. Interested readers should refer to the following before conducting a focus group (Morgan, 1993; Morgan, 1997; Krueger, 1998; Bloor, 2001). When writing items the following guidelines should be used (Spector, 1992). Items should express only one idea to minimize difficulty in responding. Both positive and negative items should be used to minimize response bias. Colloquialisms and jargon should be avoided to ensure that all respondents can comprehend the items. Items should be appropriate to the reading or comprehension level of respondents. Negatives should be avoided. For example, a phrase such as "I am not stressed often" can be confusing. Respondents often miss the negative and this can lead to misinterpretation of an item (Spector, 1992). Several texts provide additional guidance on properly wording items (Converse and Presser, 1986; Fowler, 1993, 1995; DeVellis, 2003). Multiple items provide better psychometric properties than a single item. Multi-item scales allow an increased distribution of scores, can help avoid problems associated with individual item bias and allow for higher reliability and correlation. Finally, most stress concepts are complex and multidimensional. Multiple items allow the researcher to include all dimensions of a construct (Netemeyer *et al.*, 2003). Investigators should write more items than they expect to need for the final scale (Netemeyer *et al.*, 2003). This allows the researcher to choose only the best items and drop items which are poorly worded or confusing during pre-testing and to eliminate those which do not correlate well with other items in item analysis and validity testing.

Before testing the scale, it should be translated into the proper language if being used by non-English speakers. Typically, the scale is then back-translated by another translator into English to ensure the accuracy of translation. It is important to further check the translation in the location of interest to ensure that the dialect and idioms are appropriate to the research context.

The next step in the development of a scale is pre-testing. Prior to pre-testing it is advisable to have a colleague and/or a key informant read the scale for face validity (described below), clarity and comprehensiveness. This allows a more efficient pre-test. Pre-testing can be conducted in two ways: cognitive pre-testing or field pre-testing. Cognitive pre-testing consists of a series of think-out-loud interviews. As the respondents answer the questions, the interviewer asks them to describe their thought processes in choosing the answers that they provide. Cognitive pre-testing allows the interviewer to get a good idea of the evaluation process of participants; however, participants vary in their ability to describe their cognitive processing (Fowler, 1995). Those with low education levels may find this task particularly difficult (Fowler, 1995). Another option, which achieves the same goal is to ask targeted questions which will lead you through the thought process. Fowler (1995) describes this methodology in detail and provides sample questions. Alternatively, you can conduct a field pre-test in which you conduct the interview as you would for the actual study. In this case the interviewer notes the questions which appeared to be confusing or duplicative (based on participants' comments). If you are not conducting these interviews yourself, you will want the interviewers to take notes and have a debriefing session after the interviews are completed. Fowler (1993) proposes two systematic ways of evaluating pre-tests that are more time consuming but may lead to better evaluation. The first involves creating a rating sheet for interviews to evaluate each question based on ease of reading, participant understanding and ability of respondents to answer accurately. The other option is to tape-record the pretest interviews and have trained coders code the interviews to evaluate problems. Three indications of problems with survey questions are: 1) the interviewer fails to read the question as written, 2) the respondent asks for clarification and 3) the respondent gives an inadequate answer that requires the interviewer to further probe (Oksenberg *et al.*, 1991; Fowler, 1993). There is no set number of interviews required for the pre-testing; clearly the greater the number and the more representative the sample, the better information you have. Fifteen to 50 interviews are typical for a field pre-test (Fowler, 1993, 1995). As an alternative to individual pre-testing, focus groups can also be used to evaluate scale questions (Fowler, 1995).

The next step to scale construction is establishing validity and reliability (referred to as psychometric testing). This is a multi-step process. Reliability refers to the agreement between repetitive measures of a variable. Validity refers to the agreement between the measured value and the true value. More generally, since in many cases the "true value" is never known, a valid scale is one that measures what it is supposed. Messick (1995) notes that "validity is not a property of a test or assessment as such, but rather of the meaning of tests scores" (p. 741). It is important to note that both reliability and validity are person and context specific and that assessment of reliability and validity is an ongoing process (Spector, 1992; Messick, 1995) as described below. There are several types of reliability and validity that are important in scale development.

Reliability

Inter-rater reliability refers to the agreement between different raters or, in the case of interviews, between interviewers. Inter-rater reliability can simply be examined by having two interviewers mark the questionnaire during the same set of interviews. After the interviews, the agreement is examined. A simple examination of percent agreement between raters is problematic because there will be a certain amount of agreement due to chance. The Kappa statistic is typically used to examine inter-rater reliability. Kappa is a *chance adjusted* measure of agreement.

$$\kappa = \frac{\text{observed frequency of agreement} - \text{expected frequency of agreement}}{\text{Total observed} - \text{expected frequency of agreement}}$$

Scores range from 0 to 1, with 1 indicating perfect agreement and 0 meaning no agreement. Generally, a $\kappa > 0.75$ indicates excellent agreement, $0.4 \leq \kappa \leq 0.75$ indicates good agreement and $0 \leq \kappa \leq 0.4$ indicates poor or marginal agreement (Lilienfeld and Stolley, 1994; Rosner, 1995). The Kappa statistic can be calculated with standard statistical software.

Test—retest reliability refers to agreement across time. Test—retest reliability is not appropriate for constructs that are expected to vary over time, such as mood and perceived stress (Carmines and Zeller, 1979; Spector, 1992). To calculate test—retest reliability, a similar correlation coefficient can be calculated between scale scores across time.

Internal consistency reliability is a measure of how well items of a scale measure the same construct. Internal consistency is measured during the item analysis stage of scale development. There are two primary ways of examining internal consistency, split-half analysis and coefficient (or Cronbach's) alpha. With split-half analysis the score on half of the items is compared to the score on the other half of the items with a correlation coefficient. There are numerous ways to split the scale, evens—odds, first-half—second-half or random halves (DeVellis, 2003). Split-halves is not used frequently as the results can vary depending on the way the scale is divided (Carmines and Zeller, 1979; Netemeyer *et al.*, 2003). More commonly, investigators calculate coefficient alpha. Coefficient alpha examines the common variance among items. Alpha is a measure of the portion of the scale's variance from a common source (or the construct being measured) (Netemeyer *et al.*, 2003). Coefficient alpha is a function of the number of items in the scale and the degree of intercorrelation (Carmines and Zeller, 1979; Spector, 1992). Alpha ranges from 0 to 1; scores ≥ 0.70 are considered to have good internal consistency (Spector, 1992). Kuder-Richardson formula number 20 (KR 20) is the equivalent of coefficient alpha and is used in cases when items are dichotomous (Carmines and Zeller, 1979; Spector, 1992). Both calculations can be made using standard statistical software.

Factor analysis can be used to confirm the results of item analysis and specifically to examine if the scale represents one or more constructs. A high alpha score does not guarantee that a scale represents a single construct (DeVellis, 2003). In order to assess reliability and validity, the scale must be unidimensional (Carmines and Zeller, 1979; Netemeyer *et al.*, 2003). If the scale is multidimensional, then you will want to conduct item analysis and examination of validity for each subscale separately. Dimensionality of the scale can be assessed with an exploratory factor analysis (Netemeyer *et al.*, 2003). Factor analysis allows the determination of how many factors or constructs underlie a set of items (DeVellis, 2003). Simply described, factor analysis is a group of statistical methods to examine inter-relatedness of items. The factors identified by the analysis essentially describe which cluster of items are more closely related to each other as opposed to other items in a scale. The factor loading provides an estimate of how closely an item is related to a factor; in other words, the higher the loading, the more an item contributes to the factor (Carmines and Zeller, 1979). Factor analyses of dichotomous data may not be appropriate, however, because items frequently endorsed are more likely to form factors than less commonly endorsed items even when the items represent a single domain (Nunnaly and Bernstein, 1994).

Validity

There are several different types of validity, which are categorized in several ways by different authors (Ghiselli *et al.*, 1981; DeVellis, 2003). Content validity and face validity are often used interchangeably but some argue that they are different sub-types of translational validity (Trochim, 2000; Netemeyer *et al.*, 2003). Face validity is simply a qualitative analysis to determine if the scale appears to measure what it is designed to measure. To evaluate face validity, the researcher asks experts in the field or key informants to examine the scale and assess whether the scale is hitting the mark. This is often done prior to psychometric testing but Netemeyer *et al.* (2003) suggest that it is most appropriately conducted after psychometric testing. Because of the difficulty in assessing whether a scale seems to be measuring what it is supposed to measure, face validity has been questioned as a useful approach to validity (DeVellis, 2003). Clearly, face validity is not an adequate measure of validity and should only be used as one step in the validation process. It can, however, help to improve your scale before formal testing.

Content validity is an assessment of the adequacy of items in the scale. In other words, does the scale contain all items relevant to the construct? This is easy to determine when there is a defined universe of items but it is more difficult with a complex and elusive concept such as perceived stress. Content validity is determined by having an external evaluation by experts as to the comprehensiveness of the scale. The scale should also be evaluated for irrelevant items (Messick, 1995). This is essentially a qualitative judgment; however, Netemeyer *et al.* (2003) and DeVellis (2003) recommend a procedure by which a group of judges is asked to rate scale items for appropriateness. To ensure content validity, the construct needs to be clearly defined. Additionally, the initial item pool should be chosen such that all aspects of a construct are sampled. To ensure this, items should be chosen from a variety of sources (Netemeyer *et al.*, 2003).

Criterion validity refers to the association of the scale with some external criteria. In medical research, validity is often measured against a "gold standard." The problem with assessing validity in complex constructs such as perceived stress or emotional and behavioral responses to stress is that there are truly no gold standards and the "true" values are elusive as they likely change over time. Researchers may take an analogous approach by testing a new scale against a frequently used or well-tested scale; however, this is problematic if that scale has not been well-tested in the population of interest. Lack of agreement between the scales may be due to poor validity of either the new scale or the older scale. Criterion validity can be evaluated by examining the association between the scale and any appropriate external criteria. There does not have to be a causal association between the two variables or a theoretical basis for the association to establish criterion validity (DeVellis, 2003); however, criterion validity is based on hypotheses of the direction of association (Spector, 1992). There are several subtypes of criterion validity, which are simply different approaches to assessing criterion validity (DeVellis, 2003). Concurrent validity is measured by examining the association between two variables measured at the same time. Predictive validity refers to the ability to predict some subsequently measured criteria. Concurrent and predictive validity are typically assessed by correlation or regression analyses. Known-groups or contrasting-group validity refers to the ability of the scale to distinguish between groups or the extent to which the score differs across groups as predicted. Known-groups validity is assessed by performing a t-test or an analysis of variance. Finally, convergent and discriminant validity refers to the strength of association between the scale and related and unrelated constructs. A scale has convergent validity if it has a strong relationship with a related construct and it has discriminant validity if it is unrelated or modestly related to an unrelated construct. Convergent and discriminant validity are evaluated simultaneously using a multi-trait, multi-method matrix. Convergent validity can also be assessed using factor analysis; in this case, items from the new measure will load highly on the same factor as the items from the old measure (Netemeyer *et al.*, 2003).

Construct validity refers to the ability of a scale to measure what it is supposed to measure, based on hypothesized relationships between the scale and other measures. These hypothesized relationships are based on theory. For example, stress theory suggests that perceived stress is associated with increasing blood pressure. Therefore, the scale should be empirically associated with blood pressure. Construct validity is tested in the same manner as criterion validity; the difference between them lies in the intent of the investigator (DeVellis, 2003). There is no cut point for the level of association required for construct validation. Furthermore, DeVellis (2003) points out that some association between measures may be due to the method of measurement rather than construct similarity. Multi-trait multi-method analysis, which partitions the amount of common variance into variance due to measurement similarity and variance due to construct similarity, is useful in evaluating hypothesized relationships between variables measured in a similar manner. If the scale fails to behave in a manner as theorized, it does not always indicate poor construct validity (Carmines and Zeller, 1979). Lack of association may be due to an error in the theory, error in the method used to test the association, or unreliability of the corresponding construct (Carmines and Zeller, 1979). For example, if one measures perceived stress and fails to find an association with casual blood pressure it may be due more to the variance in blood pressure than a poor measure of perceived stress. The scale could be further evaluated by measuring perceived stress while measuring blood pressure over time with an ambulatory monitor. A further complication with many physiological stress markers is that there is evidence that certain populations (very young and very old) have lower physiological reactivity levels even when they demonstrate visible distress (see Chapter 5).

Interviewer training

Once an interview protocol is developed, a critical step to a valid interview process is training the interviewers. Interviewers play a key role in the quality of the data collected. The first role of the interviewer is to put the interviewee at ease and to create an environment in which they feel comfortable to answer truthfully and accurately. One way to do this is to have a standard introduction to the interview that clearly states the purpose of the interview and the "job" of the interviewee (Fowler, 1993). In addition, interviewers should take their time reading the interview questions so that the participant feels comfortable taking the time to answer accurately (Fowler, 1993). During the consent process it should also be clearly stated that participants have the option of requesting an alternative interviewer. As human biologists often work in small, close-knit communities, interviewers and interviewees may know each other. To ensure the accuracy of the interview and to minimize participant risk, this option must be made clear. Interviewers should be encouraged to act professionally but to be personable (Fowler, 1993).

When conducting the interview, interviewers should read the question exactly as it is written. Variations in wording can introduce error in response. The interview protocol should include probes to be used when the respondent does not answer the question fully. These probes should be standardized and should be neutral. This way the interviewers do not probe in a way that is leading. Interviewers should be instructed not to assume the answer or interpret the answer. They should write responses to open-ended questions verbatim. It is advisable to have the responses written in the language used by the respondent to avoid variations in translation and interpretation. For close-ended questions, interviewers should be instructed to record the answer only when the respondent provides an appropriate answer choice. If a participant answers with a response outside of the written choices, the interviewer can repeat the question and set of answers and ask the participant to choose from one of the given answers. Finally, while the respondent should be made to feel at ease, the interviewer should remain neutral to the responses given. Interviewers should not comment on the answers provided or praise interviewees for their responses. This will minimize the motivation for providing socially desirable answers.

Fowler (1993) recommends that interview training should be conducted over 2–5 days and should include a mix of written material, presentations and role-playing. It is advisable to provide all interviewers with the research protocol book and explicit instructions on the interview procedure. The researcher should also give the field staff a clear understanding of the purpose, goals and sponsorship of the project. While providing the specific hypotheses may bias the interviewers, a clear understanding of what will be done with the data will help them better understand the purpose of the project. Interviewers should be clearly instructed on the importance of confidentiality. Finally, interviewers should practice conducting interviews with each other and with a practice group of volunteers similar to participants of interest. Training does not end with the training session. Interviewers should be directly monitored during the interview process and should be continuously reminded of the rules of interviewing.

Modes of collecting interview data

Interviews have commonly been recorded using paper and pencil. There are now several computer programs which enable the interviewer to record interviewers with a laptop or handheld computer. These programs can help improve the data collection procedure, minimize errors and significantly decrease data processing time. Computerized collection is particularly useful for interviews involving complex skip patterns, as they automatically guide the interviewer to the appropriate question. Gravlee (2002) has written a nice overview of the use of computerized data collection with a focus on Entryware software. The author has used Entryware software and found it to be extremely useful. If you choose to use such software, it is advisable to have more PDAs than needed as they may break or batteries may run low while collecting data. An on-site laptop for data back-up is critical to minimize data loss. In addition, a power source for recharging in the field is advisable (e.g. a solar panel or generator).

Diary methods

Diary data have been used in health and psychological research for many years and have proven to be useful in the study of stress (see Verbrugge, 1980; Stone *et al.*, 1991; Wheeler and Reis, 1991; Eckenrode and Bolger, 1997; Stone *et al.*, 1999) for good reviews of the history of diaries). Diary data are collected for the main purpose of providing an ecologically valid tool of everyday experiences. They have the primary advantage over interview-derived data of being less prone to recall bias. They also allow the examination of the impact of different microenvironments and real-life responses to everyday acute stressors (James, 2001). Diaries can be designed in a number of different ways and investigators can collect data on a variety of stress-related variables including stressors, appraisal of stressors, affective response, behavioral response, coping behaviors and even health outcomes. They are particularly well suited for interpreting repeated-measures physiological markers of stress but are often used independently to examine the cognitive processing of stressors. There are many designs but they generally fall into three categories: interval-contingent recording, signal-contingent recording and event-contingent recording (Wheeler and Reis, 1991; Eckenrode and Bolger, 1997).

Diary method

With interval-contingent recording, participants are instructed to record in their diaries at set intervals. The intervals can be as short as hourly or can extend up to a day. The respondent typically reports on events or feelings since the last interval. For example, Ice *et al.* (2004) asked participants to record their emotions and stressful experiences at two-hour intervals to examine the relationship between cortisol and emotions and stress. At each saliva collection time they recorded their mood, their activities in the preceding two hours and any stressful events which occurred in the preceding two hours. Mood was assessed using a modified version of the Philadelphia Geriatric Center Positive and Negative Affect Scale (Lawton *et al.*, 1992) and they were asked to describe any stressful events in an open-ended question. With frequent-interval recording, when time of measurement is critical, it is wise to use a signal such as a watch to remind the participant to record at the pre-set interval. A drawback of this design is that it is still subject to recall bias (Bolger *et al.*, 2003), although bias may be less than when participants are being asked to recall events or summarize emotions over weeks and months. In addition, bias may occur if participants are re-organizing their days around the intervals.

Signal-contingent diaries are diaries that are recorded at random intervals which are signaled by a timer, watch or personal data assistant (PDA). Two major labels have been used for signal-contingent diaries, Experience Sampling Method (ESM) (Csikszenthmihalyi *et al.*, 1977; Brandstatter, 1983; Csikszenthmihalyi and Larson, 1987) and Ecological Momentary Assessment (EMA) (Stone and Shiffman, 1994; Stone *et al.*, 1999). The main difference between ESM and EMA is that ESM refers to sampling of participant experiences, whereas EMA is a wider term referring to any kind of momentary sampling. EMA, therefore, would include not only participant experiences but physiological parameters such as ambulatory blood pressure or salivary cortisol. Stone *et al.* (1999) formally define EMA as having three characteristics:

> first, subjects are studied in the environment they typically inhabit .... Second, EMA depends on data collection about momentary or near-immediate states to avoid retrospective distortion of data .... Third, the typical EMA sampling strategy – many momentary collections per day – has two main purposes: to ensure reasonable characterization of phenomena, and to enable the researcher to examine fluctuations of phenomena over time.
>
> (Stone *et al.*, 1999, pp. 27–8).

They point out that we know little about the cognitive processes by which research participants summarize feelings or events over long periods of time. Beyond recall issues, people may vary on how they estimate their average emotion over a period. Some may base their average on frequency, others may average based on relatively few but intensive experiences. Furthermore, they point out that a person's state at the time of assessment may influence their recall. Schwartz *et al.* (1999) used EMA data on coping to calculate a "trait" measure. This measure was not associated with the trait measure obtained by interview, suggesting that a person's daily assessment can vary considerably from their recall.

With event-contingent recording, participants are asked to record events (and/or emotional response to such events) at the time an event occurs. For example, participants might be instructed to record every time they experience a stressful event (e.g. an argument with a spouse). Investigators must clearly define events so that a participant records appropriately. Event-contingent recording may be less prone to recall bias compared to interval-contingent recording and is more likely to pick up rare events than signal-contingent recording.

Which method is best? There are certainly advocates for different methods; however, the method chosen must be based on the hypothesis which the investigator wishes to test. Studies conducted using different methods may yield different results. For example, Stone *et al.* (1998) note that there was only modest agreement between signal-contingent data on coping versus those elicited from participants at the end of the day by an interviewer. However, constraints may be placed on the investigator based on the population with which s/he is working or available funding. Some populations or individuals may find the typical EMA complex sampling strategies difficult. For example, designs where some signals prompt the collection of diary data while other signals prompt the collection of physiological samples may be challenging for the young and old or those who are less time-oriented. Many anthropologists work with populations of low literacy who cannot keep diaries. Investigators interested in examining the stress process in such populations can record such events for participants. In the case of populations with low literacy, participants can be interviewed following a protocol similar to interval-contingent or signal-contingent diaries with the research staff filling out the forms instead of the participants. Event-contingent methods would be impossible unless the investigator spends a great deal of time with participants. Investigators must also consider participant burden. While there is no literature to suggest that one protocol is more burdensome than others, some may be easier for participants to fit into everyday life than others. Before launching any study, a pilot test of a diary protocol is recommended.

Frequency of recording

There is no magic number for the frequency of recording diary data. The frequency of recording must be determined by the goals of the study and the hypothesis. Rare events require more frequent sampling if using interval- or signal-contingent recording. Sampling must also be balanced against participant burden. If sampling is too frequent, not only will compliance likely decrease but it can become disruptive to an individual's life. If sampling becomes too disruptive, the ecological validity of the protocol decreases.

What to record

Again the nature of what is recorded will also be determined by the interests of the investigator. Participants can be asked a variety of questions in a check-list, Likert or open-ended format. Within stress research, the basic elements of a diary typically include time and affect. There are numerous affect measures which can be incorporated into diaries. Many are adapted from interview questionnaires (Smyth *et al.*, 1998; Ice *et al.*, 2004), while others simply ask the participant to rank their mood based on a list of adjectives on a Likert scale (James *et al.*, 1986; James *et al.*, 1990; van Eck and Nicholson, 1994). Most often, the list of adjectives used to assess mood includes both positive and negative affect as these have been shown to be independent factors (Dienner and Emmons, 1985) and often have a differential effect on physiological stress markers (James *et al.*, 1986; James *et al.*, 1990; Smyth *et al.*, 1998; Ice *et al.*, 2004). Stress studies also often include a measure of stress or anxiety (James *et al.*, 1986; Smyth *et al.*, 1998; Ice *et al.*, 2004) and may ask individuals to identify, describe and rank stressful events (van Eck and Nicholson, 1994; Ockenfels *et al.*, 1995; Smyth *et al.*, 1998). If the investigator wants to examine the impact of microenvironments on stress (e.g. job vs. home or different locations in a long-term care facility), they may also ask participants to identify their location at the time of diary recording (James *et al.*, 1991; 1996; Ice *et al.*, 2003). Questions on coping mechanisms can also be added (Stone *et al.*, 1998). Finally, researchers often add questions about variables that may influence physiological variables to enable statistical control of such factors. For ambulatory blood pressure studies, participants are typically asked to record their posture at the time of measurement as posture has an influence on blood pressure (James, 1991). In cortisol studies, participants typically report time of waking, exercise, smoking, caffeine, medication and food intake (van Eck and Nicholson, 1994; Ockenfels *et al.*, 1995; Smyth *et al.*, 1998; Ice *et al.*, 2004) all of which are known to influence cortisol. There are endless possibilities when it comes to diary data, any number of variables can be added assuming that they don't overburden participants. Readers should examine the following special issues to see the range of variables which have been studied using diary data: *Journal of Personality*, 1991, **59**(3); *Annals of Behavioral Medicine*, 1993, **15**(1); *Health Psychology*, 1998, **17**(1).

Mode of recording

Diaries can be recorded with pencil and paper or with handheld computers or personal digital assistants (PDAs). If the sampling protocol is complex (e.g. separation of diary recording from physiological sampling), a computer is recommended. Computers are also useful for complex diaries which have skip patterns as diary programs can be designed such that participants are automatically moved to the correct question. Previously, researchers who wanted to use PDAs had to create their own programs. There are now several commercial and freeware programs available. The author has not used any of the programs but there is a nice webpage created by Tamlin Conner (http://www2.bc.edu/~connert/esm.htm) which lists several freeware and commercial programs. Dr. Conner describes the advantages and disadvantages of each freeware program. Feldman Barrett and Barrett (2001) provide a general overview of electronic diaries, describing the pros and cons and a review of one freeware program, Experience Sampling Program, which they created. Generally, advantages of using computerized diaries include ability to track compliance, reduction of data-entry time and errors, and reduction of errors with skip patterns. Computerized diaries are not appropriate for use with populations with low literacy and lack of experience with computers. Additional disadvantages include extra time needed for programming, set-up and maintenance, expense and battery power limitations. Conner recommends that researchers purchase 10–20% more than needed for the research to allow for PDA attrition due to breaking and malfunctioning. Battery power is an important consideration. Researchers need to have access to electricity to recharge batteries or extra disposable batteries. Most PDAs run on rechargeable batteries. PDAs should be fully charged before being distributed to participants and sampling should not last longer than the average battery life. It is important to note that most PDAs do not have permanent storage so loss of battery power will result in loss of data. Researchers should remove most applications to prevent participants from running battery power down by playing with the other PDA programs (Feldman Barrett and Barrett, 2001).

Observational methods

Diary recording may not always be possible. Adjustments can be made such that investigators keep the diaries for participants with low literacy. With very young children and adults with cognitive impairments, such an approach will not work. In that case, the investigator can develop an observation protocol to record events, behaviors and facial expressions as a proxy for emotional response. Although observation will never give the investigator the emic perspective, it may be the only alternative when working with such populations.

Observation, particularly when performed in a naturalistic setting, is very time consuming and expensive and requires an extensive amount of training. Behavioral observation can be collected in a laboratory or naturalistic/field setting. By and large, human biologists collect data in naturalistic settings, which allows more realistic data collection but also limits the data collection approaches which can be employed.

There are numerous considerations when deciding on an observation protocol including: coding technique, types of measures (latency, frequency, duration, or intensity), sampling procedure (continuous vs. interval), collection technique (quantitative vs. qualitative), and method of analysis.

Coding

When planning an observational study, one needs to develop an "ethogram" or a catalog of behaviors/facial expressions that describe the typical behaviors of a group (Martin and Bateson, 1993). There are two approaches to coding: "social scheme" (functional scheme, e.g. sad, angry) or "physical scheme" (empirical description of expressions, e.g. forehead wrinkled) (Lerner, 1979; Bakeman and Gottman, 1997).

Both schemes require inference. Social schemes require inference before analysis; physical schemes require interpretation during or after analysis. For coding purposes, you should not be especially concerned with being truly objective but more that the observations are reproducible (i.e. good inter-observer reliability). Codes should be distinct, or precisely defined, homogenous and simple to identify but detailed enough to answer the question. In some cases you may want mutually exclusive categories but this is not always necessary. However, non-mutually exclusive categories are more of a challenge to analyze. When deciding on a coding scheme you may use a splitting or lumping approach. In some circumstances, you may want a very detailed coding scheme. In other cases, you may want more inclusive categories for analysis. One may want to use a more detailed scheme as a first step and group behaviors or expressions into larger categories during analysis. To develop a coding scheme, it is important to do some initial unstructured observation. The coding scheme should be tested prior to the beginning of the study or one should not count the first few observations in analysis to avoid "drifting" or changing of the scheme after research has started (Martin and Bateson, 1993).

Measurement types

There are several kinds of measures one can collect. The measure of interest will determine sampling method.

1. *Latency:* the time from a specified event to onset of behavior.
2. *Frequency:* the number of occurrences per a specified amount of time.
3. *Duration:* the length of behaviors.
4. *Intensity:* there is no universal definition of intensity, one must make judgments of what constitutes important components of intensity (typically people use a scale of $1-n$, e.g. the number of acts per unit time performing a discrete activity such as the number of times yelling over five-minute intervals).
5. *Sequence:* the sequence of behaviors; sequence can be recorded as real time or as the pattern of behaviors.

Sampling procedure

Once the researcher determines the measure of interest a sampling strategy must be chosen. There are two important components of sampling: who is being sampled and how the data are collected. There are several general approaches, as follows:

1. *Ad libitum*: There are no systematic constraints on when or what is recorded. You can follow one individual or a group. Generally, one records behaviors that are most obvious. Data collected in this manner cannot be quantified (Lerner, 1979).
2. *Focal person*: One individual is followed at a time (continuous or specified time intervals). This is the most common approach in human behavioral research.
3. *Scan sampling*: At specified time intervals, a group of individuals are scanned and the behavior of each individual is recorded. This approach may be particularly useful in research on social interactions or group-based interventions.
4. *Combined approach*: Simultaneous focal person and scan sampling. This technique allows examination of interactions and possible influence of others on the behavior of the individual.
5. *Behavior sampling (all occurrence sampling, event recording)*: Record each observed occurrence of a particular behavior (with or without regard to the actor).

Observations can be initiated by a timed interval or the occurrence of an event. Sampling is generally divided into continuous event recording versus timed sampling. With event recording, an observer records the time and duration (onset/offset) of a behavior or expression of interest. This method allows the determination of true frequencies and duration. Without the aid of a computer, continuous event sampling is extremely difficult to conduct because maintaining the start and stop of a variety of behaviors simultaneously with a stopwatch is challenging. With timed sampling, on the other hand, observations are based on a specified time interval; thus, duration and frequencies are estimated. The accuracy of such estimates will depend on procedure, particularly the size of time intervals. A timed sampling procedure is far easier to conduct with a paper and pencil as only a timer is need to cue an observation which can easily be recorded on a check sheet or table.

There are two basic methods of timed sampling: instantaneous and one–zero. With both, one collects data at specified time intervals. With instantaneous sampling, the behavior of interest is coded at a point in time (is it occurring or not, or what behavior is occurring). With one–zero sampling, the observer records if a behavior occurred within an interval of interest. For example, if a time interval is five minutes, the observer records whether an event has occurred within the five minutes.

The frequency within the time interval is not recorded; thus if a behavior occurred 10 times it is recorded the same as if it had only occurred once. Instantaneous sampling produces the proportion of sample points when a behavior is occurring. Continuous sampling allows the analysis of latency, frequency, duration and sequence of behaviors. Timed sampling can only provide estimates of frequency and does not allow for the analysis of latency, duration or sequence. Continuous sampling, however, is tiring for observers and may result in greater error.

Mode of recording

Observational data can be collected with pencil and paper, computer or PDA. For continuous sampling, computer or PDA recording is the preferred method to reduce observer burden and increase accuracy. When using a timed sampling procedure any recording method can be used. Pencil and paper are preferred when one does not have a specified ethogram (e.g. data on activities) or when there are an infinite number of states. Computerized data recording reduces the amount of time needed to analyze data since the data can be loaded directly into a spreadsheet or database program.

One of the biggest challenges to observational data is collecting reliable data. As the complexity of the sampling and coding scheme increases so does the time needed for training and establishment of reliability. Calculating inter-rater reliability is a tedious and time-consuming task when observational data are collected by pencil and paper. The ability to conduct reliability analysis with ease is one of the greatest advantages of computer-assisted data collection. Inter-rater reliability can be calculated quickly and therefore can assist in training.

Several software options currently exist for observational data collection, while they differ tremendously in design, flexibility and power. For those without computer programming skills, the most widely available programs are produced by Noldus Information Technology (Observer software), Education Consulting, Inc. (BEST software) and Psycsoft (Systematic Observation Software). There are also a number of producers of proprietary software (Kahng and Iwata, 1998). With one notable exception, the proprietary programs are primarily used by the creators, are difficult to get information on and have minimal technical support. MOOSES software, created by Jon Tapp and Joe Wehby at Vanderbilt University, has been widely used, particularly by education and disabilities researchers. While technical support is provided at cost, the software itself is free and available on the Internet. The above listed software are all designed for observer-based (rather than participant-based) quantitative behavioral observation. They are not appropriate for use of participant diary data nor are they equipped for qualitative observational data. With the exception of the Observer Software from Noldus, this software has primarily been designed by educational researchers and thus has some limitations in flexibility for anthropological and ethological research. For reviews of the various software options see (Kahng and Iwata, 1998; Ice, 2004; Sidener *et al.*, 2004). The author only has experience with the Observer software.

The Observer software is a flexible program which allows a researcher to develop an observational protocol, called a "configuration," using a variety of sampling strategies. The Observer is a menu-driven, Windows-based software. Noldus has made continuous improvements in their software. Observer 4.1 provided significant improvement over 3.0, a large overhaul in user interface and enhancement in documentation. The primary improvements from 4.1 to 5.0 are related to video-based observation; however, there are new features in the data collection protocol (configuration), data analysis and back-up procedure. In addition, there is a new feature, a report generator, which provides a detailed description of the protocol, data files, and analysis summaries. The Observer comes as a base package (Observer Basic) for a desktop computer with the design and analysis modules as well as an observational module. In addition, the Observer Mobile is used with handheld devices. Observer 5.0 requires at minimum Pentium II (266 MHz) with 32 MB of RAM and at least 50 MB of free space. It runs on Windows 98, Windows 2000 and Windows XP. To protect the license, Noldus has a hardware key which is fitted into a printer or USB port. The software comes on a CD-ROM. Academic prices range from $2,000 for The Observer Basic software to $5,500 for The Observer Video-Pro system.[1] Additional fees are required for use on multiple computers. For a detailed review of the Observer software and instructions on developing a protocol see Ice (2004).

Other considerations for observational data

In addition to the issues outlined above, several other factors should be considered when designing observational research. Error can occur for several reasons including observer errors due to apprehending, observer affect, errors in recording and computational error (Lerner, 1979).

One needs to consider carefully if some times are inappropriate to observe people, such as when a participant is bathing, in the bathroom, or receiving medical care. Many may view observation during these times as a violation of privacy. One has to consider the best way to analyze out-of-sight observations. Lastly, one has to decide how long each participant should be observed and how many times. Financial and time budgets are likely to dictate this issue; however, hypotheses and goals must also be weighed.

Finally, observational methods require extensive training (1–3 months). Investigators should ensure that good inter-rater reliability has been established prior to actual data collection. Training can be conducted using video recordings of behavior and/or in the field. To ensure that inter-rater reliability remains stable, approximately 10–20% of observation periods should be coded by a second observer.

Conclusions

We are still largely lacking instruments to measure stressors and emotional and behavioral responses that are valid outside of a Western context. Human biologists are in the position to greatly add to stress research and theory by developing culturally appropriate methods. All of the methods described in this chapter (interview, diary and observation) are simple to conduct; however, they require time for development, pre-testing and evaluation. Unlike physiological stress markers, the results using these methods can be seriously impacted by the population being studied, therefore investigators should not just choose a method off the shelf. Pre-testing is particularly important when working in contexts in which previously developed protocols have not been tested. With all methods, regardless of use in prior studies, reliability and validity should be re-examined.

Resource list

Education Consulting, Inc. (BEST software)
Skyware
Phone: 415 200-4519
Email: sales@skware.com
http://www.skware.com/id27.htm

Noldus Informational Technology (The Observer Software)
751 Miller Drive, Suite E-5 Leesburg, VA 20175-8993.
Phone 703-771-0440
Toll-free: 800-355-9541
Fax: 703-771-0441
Email: info@noldus.com
(Regional offices around the world can be found on the company website)
http://www.noldus.com/site/nav10000

Psychosoft (Systematic Observational Software)
Email: psychsoft@aol.com
http://members.aol.com/Psychsoft/index.html

Entryware Software
Available through SPSS
SPSS Inc.
233 S. Wacker Drive, 11th floor
Chicago, IL 60606-6307.
Phone: 312-651-3000
Fax: 312-651-3668
Email: sales@spss.com
http://www.spss.com
non-US listings: http://www.spss.com/worldwide/

Note

1 Note that these prices apply to North America only, and are subject to change (e.g. when the exchange rate changes).

References

Bakeman, R. and Gottman, J. (1997). *Observing Interaction*. Cambridge: Cambridge University Press.

Bloor, M. (2001). *Focus Groups in Social Research*. Thousand Oaks, CA: Sage.

Bolger, N., Davis, A. and Rafaeli, E. (2003). Diary methods: capturing life as it is lived. *Annual Review of Psychology*, **54**, 579–616.

Brandstatter, H. (1983). Emotional responses to other persons in everyday life situations. *Journal of Personality and Social Behavior*, **45**, 871–83.

Breslau, J. (2004). Cultures of trauma: anthropological views of posttraumatic stress disorder in international health. *Culture, Medicine and Psychiatry*, **28**, 113–26.

Carmines, E. G. and Zeller, R. A. (1979). *Reliabilty and Validity Assessment.* Newbury Park, CA: Sage.

Carver, C. S., Scheier, M. F. and Weintraub, W. S. (1989). Assessing coping strategies: A theoretically based approach. *Journal of Personality and Social Psychology*, **2**, 267–83.

Chang, E. C. (1998). Dispositional optimism and primary and secondary appraisal of a stressor: controlling for confounding influences and relations to coping and psychological and physical adjustment. *Journal of Personality and Social Psychology*, **74**(4), 1109–20.

Clark, M. S. (1992). *Emotion.* Newbury Park, CA: Sage.

Cohen, F. (1991). Measurement of coping. In *Stress and Coping: An Anthology*, ed. A. Monat and R. S. Lazarus. New York: Columbia University Press, 228–44.

  (1995). Psychological stress and susceptibility to upper respiratory infections. *American Journal of Respiratory and Critical Care Medicine*, **152**(4 Pt 2), S53–8.

Cohen, S., Doyle, W. J. and Skoner, D. P. (1999). Psychological stress, cytokine production, and severity of upper respiratory illness. *Psychosomatic Medicine*, **61**(2), 175–80.

Cohen, S., Kamarck, T. and Mermelstein, R. (1983). A global measure of perceived stress. *Journal of Health, Society and Behavior*, **24**(4), 385–96.

Cohen, S., Kessler, R. C. and Underwood, L. G. (1997). *Measuring Stress. A Guide for Health and Social Scientists.* Oxford: Oxford University Press.

Cohen, S. and Lichtenstein, E. (1990). Perceived stress, quitting smoking, and smoking relapse. *Health Psychology*, **9**(4), 466–78.

Cohen, S., Tyrrell, D. A. J. and Smith, A. H. (1993). Negative life events, perceived stress, negative affect, and susceptibility to the common cold. *Journal of Personality and Social Behavior*, **64**(1), 131–40.

Cohen, S. and Williamson, G. M. (1988). Perceived stress in a probability sample of the United States. In *The Social Psychology of Health.*, ed. S. Spacapan and S. Oskamp. Newbury Park, CA: Sage.

Converse, J. M. and Presser, S. C. (1986). *Survey Questions. Handcrafting the Standardized Questionnaire.* Newbury Park: Sage Publications.

Csikszenthmihalyi, M. and Larson, R. (1987). Validity and reliability of the experience-sampling method. *Journal of Nervous and Mental Disease*, **175**(9), 526–36.

Csikszenthmihalyi, M., Larson, R. and Prescott, S. (1977). The ecology of adolescent activity and experience. *Journal of Youth and Adolescence*, **6**, 281–94.

de Munck, V. C. and Sobo, E. J. (1998). *Using Methods in the Field. A Practical Introduction and Casebook.* Walnut Creek, CA: AltaMira Press.

Derluyn, I., Broekaert, E., Schuyten, G. and De Temmerman, E. (2004). Post-traumatic stress in former Ugandan child soldiers. *Lancet*, **363**, 861–1648.

DeVellis, R. F. (2003). *Scale Development. Theory and Applications.* Thousand Oaks, CA: Sage.

Dewe, P. (1991). Primary appraisal, secondary appraisal and coping: Their role in stressful work encounters. *Journal of Occupational Psychology*, **64**, 331–51.

Dienner, E. and Emmons, A. (1985). The independence of positive and negative affect. *Journal of Personality and Social Psychology*, **47**, 1105–17.

Eckenrode, J. and Bolger, N. (1997). Daily and within-day event measurement. In *Measuring Stress. A Guide for Health and Social Scientists*, ed. S. Cohen, R. C. Kessler and L. G. Underwood. New York: Oxford University Press.

Edgerton, R. B. (1966). Conceptions of psychosis in four East African societies. *American Anthropologist*, **68**(2), 408–25.

(1971). A traditional African psychiatrist. *Southwestern Journal of Anthropology*, **27**, 259–78.

(1980). Traditional treatment for mental illness in Africa: A review. *Culture, Medicine and Psychiatry*, **4**, 167–89.

Edgerton, R. B. and Cohen, A. (1994). Culture and schizophrenia: The DOSMD challenge. *British Journal of Psychiatry*, **164**, 222–31.

Feldman Barrett, L. and Barrett, D. J. (2001). An introduction to computerized experience sampling in psychology. *Social Science Computer Review*, **19**(2), 175–85.

Fliege, H., Rose, M., Arck, P. *et al.* (2005). The Perceived Stress Questionnaire (PSQ) reconsidered: validation and reference values from different clinical and healthy adult samples. *Psychosomatic Medicine*, **67**, 78–88.

Folkman, S. and Lazarus, R. S. (1980). *Manual for the Ways of Coping Questionnaire*. Palo Alto, CA: Consulting Psychologists Press.

Folkman, S., Lazarus, R. S., Gruen, R. J. and DeLongis, A. (1986). Appraisal, coping, health status and psychological symptoms. *Journal of Personality and Social Psychology*, **50**, 571–9.

Fowler, F. J. (1993). *Survey Research Methods*. Newbury Park, CA: Sage.

(1995). *Improving Survey Questions. Design and Evaluation*. Thousand Oaks, CA: Sage.

Ghiselli, E. E., Campbell, J. P. and Zedeck, S. (1981). *Measurement Theory for the Behavioral Sciences*. San Francisco, CA: W.H. Freeman and Company.

Gravlee, C. E. (2002). Mobile computer-assisted personal interviewing with handheld computers: the Entryware System 3.0. *Field Methods*, **14**(3), 322–36.

Hewitt, P. L., Flett, G. L. and Mosher, S. W. (1992). The perceived stress scale: factor structure and relation to depression symptoms in a psychiatric sample. *Journal of Psychopathology and Behavioral Assessment*, **14**(3), 247–57.

Ice, G. H. (2004). Technological advances in observational data collection: the advantages and limitations of computer assisted data collection. *Field Methods*, **16**(3), 352–75.

Ice, G. H., James, G. D. and Crews, D. (2003). Blood pressure variation in the institutionalized elderly. *Collegium Antropologicum*, **27**(2), 47–56.

Ice, G. H., Katz-Stein, A., Himes, J. H. and Kane, R. L. (2004). Diurnal cycles of salivary cortisol in older adults. *Psychoneuroendocrinology*, **29**(3), 355–70.

James, G. D. (1991). Blood pressure response to daily stressors of urban environments: methodology, basic concepts and significance. *Yearbook of Physical Anthropology*, **34**, 189−210.

  (2001). Evaluation of journals, diaries, and indexes of worksite environmental stress. In *Contemporary Cardiology: Blood Pressure Monitoring in Cardiovascular Medicine and Therapeutics*, ed. W. White. Totowa, NJ: Humana Press, pp. 29−44.

James, G. D., Broege, P. and Schlussel, Y. (1996). Assessing cardiovascular risk and stress-related blood pressure variability in young women employed in wage jobs. *American Journal of Human Biology*, **8**, 743−9.

James, G. D., Moucha, O. P. and Pickering, T. G. (1991). The normal hourly variation of blood pressure in women: average patterns and the effect of work stress. *Journal of Human Hypertension*, **5**(6), 505−9.

James, G. D., Yee, L. S., Harshfield, G. A., Blank, S. G. and Pickering, T. G. (1986). The influence of happiness, anger, and anxiety on the blood pressure of borderline hypertensives. *Psychosomatic Medicine*, **48**(7), 502−8.

James, G. D., Yee, L. S. and Pickering, T. G. (1990). Winter−summer differences in the effects of emotion, posture and place of measurement on blood pressure. *Social Science and Medicine*, **31**(11), 1213−17.

Jenkins, J. H. and Karno, M. (1992). The meaning of expressed emotion: theoretical issues raised by cross-cultural research. *American Journal of Psychiatry*, **149**, 9−21.

Kaaya, S. F., Fawzi, M. C. S., Mbwambo, J. *et al.* (2002). Validity of the Hopkins Symptom Checklist-25 amongst HIV-positive pregnant women in Tanzania. *Acta Psychiatrica Scandinavica*, **106**, 9−19.

Kahng, S. and Iwata, B. A. (1998). Computerized systems for collecting real-time observational data. *Journal of Applied Behavior Analysis*, **31**, 253−61.

Katz, M. M., Marsella, A., Dube, K. C. *et al.* (1988). On the expression of psychosis in different cultures: schizophrenia in an Indian and in a Nigerian community. *Culture, Medicine and Psychiatry*, **12**, 331−55.

Krueger, R. (1998). *Focus Group Tool Kit*. Thousand Oaks, CA: Sage.

Larsen, R. J. and Diener, E. (1992). Promises and problems with the circumplex model of emotion. In *Emotion*, ed. M. S. Clark. Newbury Park, CA: Sage, 13: 25−59.

Lawton, M. P., Kleban, M. H., Dean, J., Rajagopal, D. and Parmelee, P. A. (1992). The factorial generality of brief positive and negative affect measures. *Journal of Gerontology*, **47**(4), 228−37.

Lazarus, R. S. (1984). Puzzles in the study of daily hassles. *Journal of Behavioral Medicine*, **7**, 375−89.

  (1991). *Emotion and Adaptation*. New York: Oxford University Press.

Lazarus, R. S. and Folkman, S. (1984a). Coping and adaptation. In *Handbook of Behavioral Medicine*, ed. W. D. Gentry. New York: Guilford Press, pp. 282−325.

  (1984b). *Stress, Appraisal and Coping*. New York: Springer.

Lerner, P. (1979). *Handbook of Ethological Methods*. New York: Garland STPM Press.

Levenstein, S., Prantera, C., Varvo, V. *et al.* (1993). Development of the Perceived Stress Questionnaire: a new tool for psychosomatic research. *Journal of Psychosomatic Research*, **37**(1), 19–32.

Lilienfeld, D. E. and Stolley, P. D. (1994). *Foundations of Epidemiology*. New York: Oxford University Press.

Marlowe, N. (1998). Stressful events, appraisal, coping and recurrent headache. *Journal of Clinical Psychology*, **59**(10), 1107–16.

Martin, P. and Bateson, P. (1993). *Measuring Behavior*. Cambridge: Cambridge University Press.

Martin, R. A., Kazarian, S. S. and Breiter, H. J. (1995). Perceived stress, life events, dysfunctional attitudes, and depression in adolescent psychiatric inpatients. *Journal of Psychopathology and Behavioral Assessment*, **17**(1), 81–95.

McNair, D., Lorr, M. and Droppleman, L. (1971). *Psychiatric Outpatient Mood Scale*. Boston: Psychopharmacology Laboratory, Boston University Medical Center.

Messick, S. (1995). Validity of psychological assessment. Validation of inferences from persons' responses and performances as scientific inquiry into score meaning. *American Psychologist*, **50**(9), 741–9.

Monroe, S. M. and Kelley, J. M. (1997). Measurement of stress appraisal. In *Measuring Stress. A Guide for Health and Social Scientists.*, ed. S. Cohen, R. C. Kessler and L. U. Gordon. New York: Oxford University Press, pp. 122–47.

Morgan, C. A. (1993). *Successful Focus Groups: Advancing the State of the Art*. Newbury Park, CA: Sage.

Morgan, D. L. (1997). *Focus Groups as Qualitative Research*. Thousand Oaks, CA: Sage.

Morris, J. E. and Long, B. C. (2002). Female clerical workers' occupational stress: the role of the person and social resources, negative affectivity and stress appraisals. *Journal of Counseling Psychology*, **49**(4), 395–410.

Myers, F. R. (1979). Emotions and self: a theory of personhood and the political order among Pintupi Aborigines. *Ethos*, **7**, 343–70.

Netemeyer, R. G., Bearden, W. O. and Sharma, S. (2003). *Scaling Procedures. Issues and Applications*. Thousand Oaks, CA: Sage.

Nunnaly, J. and Bernstein, I. (1994). *Psychometric Theory*. New York: McGraw Hill.

Ockenfels, M. C., Porter, L., Smyth, J. *et al.* (1995). Effect of chronic stress associated with unemployment on salivary cortisol: overall cortisol levels, diurnal rhythm, and acute stress reactivity. *Psychosomatic Medicine*, **57**(5), 460–7.

Oksenberg, L., Cannell, C. and Kalton, G. (1991). New strategies of pretesting survey questions. *Journal of Official Statistics*, **7**(3), 349–66.

Patel, V. (2001). Cultural factors and international epidemiology. *British Medical Bulletin*, **57**, 33–45.

Patel, V., Araya, R., de Lima, M., Ludermir, A. and Todd, C. (1999). Women, poverty and common mental disorders in four restructuring societies. *Social Science and Medicine*, **49**, 1461–71.

Peacock, E. J. and Wong, P. T. P. (1990). The stress appraisal measure (SAM): a multidimensional approach to cognitive appraisal. *Stress Medicine*, **6**, 227–36.

Pham, P. N., Weinstein, H. M. and Longman, T. (2004). Trauma and PTSD symptoms in Rwanda. Implications for attitudes toward justice and reconciliation. *Journal of the American Medical Association*, **292**(5), 602–12.

Pike, I. L. and Young, A. (2002). Understanding psychosocial health among reproductive age women from Turkana District, Kenya and Mbulu District, Tanzania. *American Anthropological Association Annual Meeting Abstracts*, **479**.

Plutchik, R. (1989). Measuring emotions and their derivatives. In *Emotion: Theory, Research and Experience, Volume, 4. The Measurement of Emotions.* ed. R. Plutchik and H. Kellerman. San Diego, CA: Academic Press, pp. 1–35.

Plutchik, R. and Kellerman, H. (1989). *Emotion. Theory, Research and Experience.* San Diego, CA: Academic Press.

Rosner, B. (1995). *Fundamentals of Biostatistics.* Belmont, CA: Duxbury Press.

Rowley, A. A., Roesch, S. C., Jurica, B. J. and Vaughn, A. A. (2005). Developing and validating a stress appraisal measure for minority adolescents. *Journal of Adolescence*, **28**, 547–57.

Sanz-Carrillo, C., Garcia-Campayo, J., Rubio, A., Santed, M. A. and Montoro, M. (2002). Validation of the Spanish version of the Perceived Stress Questionnaire. *Journal of Psychosomatic Research*, **52**(3), 167–72.

Schwartz, J. E., Neale, J., Marco, C., Shiffman, S. S. and Stone, A. A. (1999). Does trait coping exist? A momentary assessment approach to the evaluation of traits. *Journal of Personality and Social Psychology*, **77**(2), 360–9.

Schwarzer, R. and Schwarzer, C. (1996). A critical survey of coping instruments. In *Handbook of Coping: Theory, Research, Applications*, ed. M. Zeidner and N. S. Endler. New York: Wiley, pp. 107–32.

Shaver, P. R., Wu, S. and Schwartz, J. C. (1992). Cross-cultural similarities and differences in emotion and its representation. In *Emotion*, ed. M. S. Clark. Newbury Park: Sage. 13: 175–212.

Sidener, T. M., Shabani, D. B. and Carr, J. E. (2004). A review of the Behavioral Evaluation Strategy & Taxonomy (BEST) Software application. *Behavioral Interventions*, **in press**.

Smyth, J., Ockenfels, M., Porter, L., Kirschbaum, C., Hellhammer, D. and Stone, A. (1998). Stressors and mood measured on a momentary basis are associated with salivary cortisol secretion. *Psychoneuroendocrinology*, **22**, 353–70.

Spector, P. E. (1992). *Summated Rating Scale Construction. An Introduction.* Newbury Park, CA: Sage.

Stone, A. (1997). Measurement of affective response. In *Measuring Stress. A Guide for Health and Social Scientists.*, ed. S. Cohen, R. C. Kessler and L. U. Gordon. New York: Oxford University Press.

Stone, A. A. and Shiffman, S. (1994). Ecological momentary assessment (EMA) in behavioral medicine. *Annals of Behavioral Medicine*, **16**, 199–202.

Stone, A. A., Kessler, R. C. and Haythornthwaite, J. A. (1991). Measuring daily events and experiences: decisions for the researcher. *Journal of Personality*, **59**, 575–605.

Stone, A. A., Schwarz, J. E., Neale, J. M. *et al.* (1998). How accurate are current coping assessments? A comparison of momentary versus end of the day coping efforts. *Journal of Personality and Social Psychology*, **74**, 1670–80.

Stone, A. A., Shiffman, S. S. and DeVries, M. (1999). Ecological momentary assessment. In *Understanding Quality of Life: Scientific Perspectives in Enjoyment and Suffering*, ed. E. D. Kahneman and N. Schwartz. New York: Russell Sage, pp. 26–39.

Trochim, W. (2000). *The Research Methods Knowledge Base*. 2nd edn. Cincinnati, OH: Atomic Dog.

van Eck, M. and Nicholson, N. (1994). Perceived stress and salivary cortisol in daily life. *Annals of Behavioral Medicine*, **16**, 221–7.

Verbrugge, L. (1980). Health Diaries. *Medical Care*, **19**(73–95).

Watson, D., Clark, L. and Tellegen, A. (1988). Development and validation of brief measure of positive and negative affect: The PNAS Scales. *Journal of Personality and Social Psychology*, **54**(6), 1063–70.

Watson, D. and Vaidya, J. (2003). Mood measurement: current status and future directions. In *Handbook of Psychology, Volume 2, Research Methods in Psychology*. Hoboken, NJ: Wiley.

Weller, S. C. and Romney, A. K. (1988). *Systematic Data Collection*. Newbury Park, CA: Sage.

Wheeler, L. and Reis, H. T. (1991). Self-recording of everyday life events: Origins, types and uses. *Journal of Personality*, **59**, 339–54.

4 *Measuring hormonal variation in the sympathetic nervous system: catecholamines*

DANIEL E. BROWN

It is commonplace for people to conceptualize the body's response to stress as an "adrenaline rush," with the symptoms of this response, as described by Cannon (1915), characterizing the "fight or flight" response. This common understanding of the stress response encompasses one major component of biological stress: activation of the sympathetic adrenal medullary system (SAMS). The other main arm of the stress response, the hypothalamic pituitary adrenal cortex (HPA) axis, will be discussed in the next chapter. Activation of the SAMS has widespread effects in the body. These effects are seen as being allostatic in nature; that is, they help maintain homeostasis through the initiation of physiological change. Measures of stress have been used that identify the indirect effects of SAMS activation such as increased pulse and blood pressure. Other measures, to be examined here, focus on more direct evidence for SAMS activation.

Stress has taken on great importance for humans in the modern world. The most common ailments that threaten us, such as cardiovascular disease, diabetes, and even some cancers, apparently are influenced by chronically high stress levels. The response that has allowed our ancestors to adapt to the stressors that were part of their world now appears too often to be an inappropriate one for the kinds of stressors, from traffic to deadlines, that confront us today (Sapolsky, 1994).

The stress response represents adaptation to actual or perceived challenges from the environment, with this process termed "allostasis" (McEwen, 1998). On the other hand, the stress response can also cause harm to the body if it is invoked too frequently or functions abnormally, with this overuse or malfunction termed "allostatic load" (McEwen and Stellar, 1993). Malfunction can be due to the stress response's failure to turn off promptly when the stress has abated, or the failure to turn on in a timely manner when the stress first appears. Allostatic load can lead to the so-called "diseases of adaptation" that result from wear and tear on the body caused by overuse or malfunction of the physiological stress response. One of the measures used to quantify allostatic load in humans is overnight catecholamine excretion rates, an indicator of chronic SAMS activity.

The SAMS, besides invoking the fight or flight response, has important functions in maintaining homeostasis during activity. For instance, the sympathetic nervous system allows changes in blood flow and blood pressure to permit adequate oxygen and energy supply to the brain and muscles during posture changes and movement (McEwen, 2003). Catecholamines released by the SAMS also mobilize energy resources, leading to increases in fat and sugar levels in the blood, control body fluid volume, and also affect immune function during stress. While these functions are absolutely necessary for survival, allostatic load character- ized by overstimulation of the SAMS can lead to hypertension, dyslipidemia, and other serious health risks for the individual.

Stress itself is usually characterized by psychologists as neither a stimulus (i.e., an environmental characteristic) nor a response (i.e. a physiological reaction), but rather a "mediating variable" that connects a stimulus with a response. The mediation occurs in the brain, and, according to a commonly used psychological theory, involves interpreta- tion of a given stimulus as a threat that cannot be coped with routinely (Lazarus, 1966) (see Chapter 3 for more information on coping).

Activation of the stress response

The interpretation of stress is believed to occur in the cerebral cortex of the brain based upon sensory and other input (such as from chemo- receptors). It has become common to differentiate two main types of stress categories: "emotional" (or "neurogenic") and "physiological" (or "systematic") (Li *et al.*, 1996; Sawchenco *et al.*, 2000). The two cate- gories of stress act through somewhat different neurological mechanisms, but the general features are similar. Cognitive processes are involved in the assessment of the input for both categories as to whether the input represents a potential threat and, if so, whether there are coping processes to deal with the threat in a routine manner. If the appraisal is of a threat that cannot be coped with easily, a stress message is sent from the cortex to other brain areas. These other areas include the amygdala, the hypothalamus, the locus coeruleus and the rostral ventrolateral medulla (RVLM) in the brain stem. These brain areas contain neurons that release norepinephrine, those that release corticotropin-releasing hormone (CRH), and those that release arginine vasopressin (AVP), with these types of neurons stimulating each other (Calogero *et al.*, 1988; Chrousos and Gold, 1992; Charmandari *et al.*, 2003).

The amygdala appears to have a central role in the stress response (Allen and Allen, 1974; Kim *et al.*, 2001). Amygdala neurons release CRH which has two major effects: it causes the brain stem (including the RVLM) to stimulate the sympathetic nervous system (SNS) through spinal nerves, and it also activates the HPA axis of the stress response. Thus, the two arms of the stress response are interconnected. The RVLM is a major controller of sympathetic tone (Goldstein, 1987; Calaresu and Yardley, 1988) and therefore seems to hold an important role in the SAMS response to stress.

The brain's stimulation of the SNS proceeds through the spinal cord. Sympathetic nerves project from the ventral roots of the thoracic and two upper lumbar spinal nerves to a series of sympathetic ganglia that form a chain paralleling both sides of the spinal column. Sympathetic pre-ganglionic fibers synapse at the ganglia, releasing the neurotransmitter acetylcholine, which in turn activates post-ganglionic fibers which project from the ganglia to many different target organs throughout the body. Some pre-ganglionic fibers synapse at sympathetic ganglia outside this chain. The post-ganglionic fibers release the neurotransmitter norepinephrine at their organ target sites (with the exception that fibers which target sweat glands release acetylcholine). There is a re-uptake of much of the norepinephrine by nerves. This re-uptake, along with methylation of norepinephrine, makes its action last only seconds. The "spillover" of norepinephrine leaks into plasma. Some pre-ganglionic fibers synapse directly with secretory cells in the adrenal medulla. These cells secrete a mixture of epinephrine (about 75 percent) and norepinephrine (about 25 percent) directly into the bloodstream. There is a slight difference in the proportions of the two hormones released from the adrenal medulla, depending in part upon what type of stimulus causes their release. For instance, hypoglycemia and nicotine increase the proportion of epinephrine, while carotid occlusion leads to lower epinephrine proportions (Lewis, 1975).

When activated by stress, the SNS does not act in a homogeneous manner. When SNS nerve activity has been monitored at several sites simultaneously, differences in the response have been noted in different circumstances, suggesting that the SNS response is not all-or-none (Ninomiya *et al.*, 1973; Dimsdale and Ziegler, 1991). Effector organs are stimulated via two major adrenergic receptors: alpha and beta, as described below. Epinephrine activates both types equally, but norepinephrine has a more pronounced effect on alpha receptors. Stimulation of the two types of receptors has differing effects. For instance, alpha receptor stimulation generally leads to vasoconstriction, while beta receptor stimulation generally leads to vasodilation. Accordingly, norepinephrine has a greater effect on blood pressure elevation than does epinephrine.

Thus, a complex series of brain pathways that involve both cognitive and autonomic areas result in the secretion of norepinephrine at nerve terminals in many organs and the release of a mixture of epinephrine and norepinephrine into the blood from the adrenal medulla. These catecholamines cause a series of physiological changes to occur in the body, including increased heart rate and blood pressure, elevated ventilation rate, released energy stores into the blood, blood preferentially routed to muscles, and so forth: the changes that categorize the "fight or flight" response. SNS activation is not an all-or-none response, as specific parts of the SNS system are under greater activation than others under certain conditions.

Effects of SNS activation

There are general effects of sympathetic activation, as listed in Table 4.1. There is a marked functional distinction between effects of norepineph-rine and epinephrine, and differential activation of sympathetic nerves under varying stimuli, so this general picture of sympathetic effects is somewhat simplified.

The catecholamines exert their effects through interactions with receptors on cell surfaces in target organs. There are three main types of receptors, alpha-1, alpha-2 and beta, but there are also subtypes (Insel, 1996). The receptors are transmembrane proteins which have either stimulatory or inhibitory effects on intracellular signaling pathways. Each of the three types of receptor reacts with a different major grouping of the transmembrane proteins, which are termed G proteins. The main receptors are:

> *Alpha-1 receptors*: These bind both to norepinephrine and epinephrine, and cause an increase in free calcium inside the cell. They are found in blood vessels, particularly in the skin and gastrointestinal system, where they cause vasoconstriction and thus decreased blood supply to these organs during stress. Three human subtypes (α_{1A}, α_{1B} and α_{1D}) have been identified

Table 4.1. *Physiological effects of sympathetic activation*

Organ	Effect of sympathetic stimulation
Eccrine glands	Increased sweating
Apocrine glands	Increased sweating
Eye (pupil)	Dilation
Skin	Decreased blood supply
Heart	Faster pulse; increased force of beat
Lungs	Dilation of bronchi; increased ventilation
Gastrointestinal	Decreased peristalsis; decreased blood supply
Liver	Release of glucagon
Pancreas	Inhibited secretion of insulin
Kidney	Decreased output; sodium retention
Penis	Ejaculation
Systemic blood vessels	Constricted (adrenergic action); dilated (cholinergic action)
Blood	Increased coagulation
Basal metabolism	Sharply increased
Adrenal cortex	Increased secretion of adrenal corticosteroids
Mental activity	Generally increased; externally directed
Piloerector muscles	Excited

Source: Guyton (1971).

and are associated with different genes (Hague *et al.*, 2003). The subtypes are also somewhat different in their characteristic location (based on studies in the rat), with α_{1A} found particularly in blood vessels, the submaxillary gland, vas deferens, and urethra; α_{1B} characteristically found in the liver and spleen; and α_{1D} found in major arteries (Piascik and Perez, 2001).

Alpha-2 receptors: These also bind both to norepinephrine and epinephrine, and cause a decrease in cyclic AMP inside the cell. They are found on pre-synaptic nerve terminals. Three subtypes of these receptors have also been found. The α_{2A} receptor is found widely throughout the CNS and peripheral areas of the body; the α_{2B} receptor is found in high levels in the kidney, as well as in other areas of the body outside the CNS; while the α_{2C} receptor is primarily located in the CNS, with small amounts in the kidney (Link *et al.*, 1996). Activation of α_{2A} receptors in the RVLM lowers sympathetic outflow, and thus tends to lower systemic blood pressure (MacMillan *et al.*, 1996). On the other hand, stimulation of alpha-2 receptors in arterial smooth muscle cells heightens vascular resistance, and thus increases blood pressure (Link *et al.*, 1996). Hence, activation of the subtypes of the alpha-2 receptors has greatly varying physiological effects.

Beta-1 receptors: These bind both to norepinephrine and epinephrine, leading to an increase in cyclic AMP within the cell. They are present in heart tissue, particularly on cells in the heart's pacemaker regions, where they cause increased heart rate. Beta-1 receptor antagonists have been utilized in therapy for people with coronary artery disease, hypertension, and chronic heart failure (Frishman and Lazar, 1990; Psaty *et al.*, 1997).

Beta-2 receptors: These bind preferentially to epinephrine rather than norepinephrine, and cause an increase in intracellular cyclic AMP. They are present in blood vessels within skeletal muscle, where they cause vasodilation, and thus an increased blood supply during stress. Beta-2 receptor agonists have been employed in asthma therapy. Beta-2 receptors on erythrocytes have also been implicated in malarial infection (Harrison *et al.*, 2003).

Beta-3 receptors: These bind to both epinephrine and norepinephrine, and stimulate an increase in intracellular AMP. They are present in adipose tissue, particularly brown adipose tissue, and may have an important role in regulating lipid metabolism (Lafantan and Berlan, 1993). Unlike beta-1 and beta-2 receptors, beta-3 receptors do not show desensitization after short-term activation (Nantel *et al.*, 1994).

It can thus be seen that the effects of catecholamines are quite variable depending on which receptors are activated and where in the body the receptors are located. There are also species differences in the prevalence and distribution of the receptors.

Catecholamines, through interactions with beta receptors, also increase the amount of natural killer and lymphokine activated killer cells in the immune system during stress reactions (Pederson and Hoffman-Goetz, 2000). This is the opposite of the effects of corticosteroids that are released during activation of the HPA axis, although the time courses are different: catecholamines have an immediate effect on immune cells, while the corticosteroids act over a longer period of time.

Measuring SAMS activity

The above description of the stress response provides landmarks for measurements of SAMS activity. The most direct measurements would consist of using microelectrodes to measure neuronal activity in stress

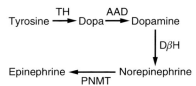

Figure 4.1.   Biochemical pathway of catecholamine synthesis (TH = tyrosine hydroxylase; AAD = aromatic acid decarboxylase; DβH = dopamine-β-hydroxylase; PNMT = phenylethanolamine N-methyltransferase).

centers of the brain. Laboratory studies of animal models utilize this procedure, although our ability to identify and measure specific CNS neurons involved in stress activation is limited. The next most direct measures involve the use of clinical microneurography to measure sympathetic nerve activation in subcutaneous sympathetic nerves outside the brain (e.g., Wilkinson, *et al.*, 1998). Human studies done in carefully controlled laboratory or hospital environments have been carried out using this form of measurement.

Other studies have analyzed the concentration of catecholamines in cerebrospinal fluid (CSF). Catecholamines are the class of molecules that includes norepinephrine and epinephrine. In fact, a biochemical pathway links the catecholamines, as shown in Figure 4.1. Norepinephrine is thought to leak into CSF from the central nervous system, and the concentration of this neurotransmitter reflects the activity of noradrenergic neuronal transmission, particularly that from the locus coeruleus and projections from that structure (Foote *et al.*, 1983; Esler, 1995). There is evidence that CSF norepinephrine is elevated in people with hypertension, suggesting a link between this stress measure and a presumed stress-related disease (Eide *et al.*, 1979). Unfortunately, obtaining CSF is an invasive procedure that elevates stress in human subjects and is not practical for the monitoring of stress levels during normal life activities.

Our purpose is not only to measure stress under laboratory conditions, but also to monitor stress levels in people who are experiencing normal life activities. It is only in this way that we can identify what circumstances generate stress in people, with possible negative consequences for health. The measurements outlined above that involve implanting of electrodes into humans or collecting cerebrospinal fluid are far too invasive for use in monitoring normal human activities. In fact, we face the prospect of actually causing the stress we are attempting to monitor by the act of the measurement. Thus, we must depend on less direct means of measuring SAMS activity.

The next most direct means of measuring SAMS activity is by assaying the concentration of catecholamines in blood samples. As noted above, norepinephrine that is not eliminated through re-uptake into nerves is "spilled over" into plasma, while epinephrine and some norepinephrine are released directly into plasma from the adrenal medulla. Studies using radioactive-labeled norepinephrine permit assessment of the appearance rates of norepinephrine in plasma, with rates measured either regionally or for the whole body (Esler, 1993). It is estimated that 10–20 percent of norepinephrine released by sympathetic nerves overflows into the blood, with the plasma norepinephrine levels proportional to the rate of sympathetic nerve firing (Esler *et al.*, 1990). Much of the spillover norepinephrine originates from the kidneys, skeletal muscle and lungs. Some norepinephrine is also released by chromaffin cells in the adrenal medulla.

Plasma levels of catecholamines also depend upon clearance rates, with these hormones removed from the plasma through neuronal uptake, re-uptake by other tissues, and metabolic conversion. The clearance occurs quickly, with values ranging from 1–3 l/min (Best and Halter, 1982; Esler *et al.*, 1990), and the half-life of plasma catecholamines is 1–2 minutes (Silverberg *et al.*, 1978; Fitzgerald *et al.*, 1979; Ward and Mefford, 1985). This leads to a marked difference in catecholamine levels between arterial and venous blood. Therefore, plasma levels of catecholamines must be considered only an indirect measure of SNS activity, although these levels are highly correlated with more direct measures of sympathetic activity, such as sympathetic nerve firing rate (Wallin *et al.*, 1981), and are also related to rates of cardiovascular pathology (Dimsdale and Ziegler, 1991).

A concern with using catecholamines in plasma or urine as a stress measure is that they measure, in an indirect manner, the total amount of SNS activity over a given amount of time. Studies have shown that the SNS does not act in a uniform manner to any stress stimulus (Ninomiya *et al.*, 1973; Esler *et al.*, 1988; Dimsdale and Ziegler, 1991). Catecholamine measurement in plasma or urine cannot differentiate between SNS activation that may be stronger to some target organs than to others.

A greater concern in the use of plasma catecholamines is that the very act of obtaining samples causes stress, and thus changes the catecholamine level that one attempts to measure. Venipuncture is a potent stimulator of SNS activity, increasing plasma catecholamine levels by over 50 percent (Carruthers *et al.*, 1970). Given the anticipatory nature of the stress reaction in humans, the sight of the investigator coming, needle in hand, will elicit a profound increase in catecholamine level before the sample can be drawn. If multiple samples are to be taken when monitoring stress, an indwelling catheter or butterfly valve may be used. Even so, this does not permit monitoring of stress in people doing daily activities. A less invasive measure is required to allow monitoring of normal stress levels encountered by people.

Catecholamines also appear in the urine, and obtaining urine samples is much less invasive than blood collection. However, because of its rapid clearance from plasma, only about 10 percent of plasma norepinephrine is deposited in the urine. As only 10–20 percent of norepinephrine from sympathetic nerves goes into plasma this means that only between 1–2 percent of the norepinephrine released by those sympathetic nerves ends up in urine. Urinary catecholamines are therefore a very indirect measure of SNS activity. Validity of urinary catecholamines as a measure of SNS activity is further questioned by the finding that the kidney can synthesize epinephrine (Ziegler *et al.*, 1989). Furthermore, increased SNS activity reduces renal blood flow, and thus slows the movement of catecholamines from plasma to urine (Dimsdale and Ziegler, 1991).

The collection requires the carrying of bottles and the proper recording of collection times by study participants, and the catecholamines represent levels secreted over a range of time, usually a matter of several hours (Ward and Mefford, 1985). Diet may have an effect on urinary catecholamine levels, as certain dietary items such as bananas increase sulfate-conjugated norepinephrine in plasma and urine (Davidson *et al.*, 1981; Dunne *et al.*, 1983). If urine is properly preserved and refrigerated, the free norepinephrine levels are not affected by these increased amounts of sulfate-conjugated catecholamines.

There are certain field situations where the collection of urine is difficult or impossible, particularly because freezing of samples should be done within 24 hours of collection. Also, there may be cultural factors which militate against the taking of urine samples.

Despite these drawbacks, urinary catecholamine levels are significantly correlated with plasma levels of these hormones (Januszewicz *et al.*, 1979; Akerstedt *et al.*, 1983), although low correlations are sometimes found when measuring over short time intervals (Euler, 1967). Urinary catecholamine measurements are therefore only useful when measuring over a period of several hours or longer, which provides a measure of response to prolonged or averaged intensity of stress as opposed to momentary events (James *et al.*, 1989). In choosing a measure of stress, investigators must therefore be cognizant of differences in time span of the stress to be measured; this time span is specific to the research design and hypothesis to be tested. Table 4.2 shows some stress measures and the duration of response that they quantify. Due to the different time spans

Table 4.2. *Time duration of selected stress measures related to the SAMS*

Measurement	Approximate time duration
SNS nerve firing	Fractions of seconds
Plasma catecholamines	Seconds–minutes
Urinary catecholamines	Hours–days
Ambulatory blood pressure	Minutes
Heart rate changes	Seconds–minutes

that are measured, correlations between these techniques are not necessarily high.

A need for standardization in the use of urinary catecholamines

A major concern over the use of urinary catecholamines as a measure of stress is in the standardization of protocols and assays. The length of time sampled has varied in studies from a few hours to 24 hours. Some variability in time is appropriate as the time span covered needs to correspond with the specific research questions being posed. However, some degree of standardization of time spans for collections should be considered. Use of 24-hour collections may be appropriate in some circumstances where chronic stress levels are desired. In many studies approximately 4-hour collections during waking hours in particular daily settings (e.g., work, home, other) and approximately 8-hour collections overnight have been utilized (James, *et al.*, 1993; Brown and James, 2000). The latter sampling strategy permits observation of the effects of different settings on stress levels.

There have been two methods used for stabilizing the catecholamines during the collection period. Some protocols involve collection of urine in acid to inhibit the oxidation of catecholamines, while others have utilized sodium metabisulfite which also inhibits oxidation (Boomsma *et al.*, 1993; Miki and Sudo, 1998). Either method stabilizes catecholamines for several months in frozen urine.

A more important concern for comparability of studies utilizing urinary excretion of catecholamines as a stress measure is the differing analytical techniques used in assays. The earliest methods, starting early in the twentieth century, involved colorimetric approaches, which were not sensitive enough to measure physiological levels of catecholamines in blood, and bioassay methods, which also were not effective (Callingham, 1975). Advances were made in the 1940s and 1950s in extraction of catecholamines, with alumina or ion-exchange resins often being employed currently for these extractions. Contemporary assay techniques include: radioenzymatic assays using catechol-O-methyltransferase (COMT), including single and double isotope assays; phenylethanolamine N-methyltransferase (PNMT) radioenzymatic assays; fluorimetric assays; and HPLC (high performance liquid chromatographic) assays, including reverse phase and cation exchange chromatography. Even when using the same one of these techniques, different laboratories sometimes use somewhat different details in the assay protocols. Hjemdahl (1984) did a study comparing catecholamine determinations in plasma from different laboratories using these various techniques, sending aliquots from two pooled samples to multiple laboratories that did catecholamine assays. He noted large discrepancies from the other methods by the fluorimetric technique, and this technique also had a considerably larger coefficient of variation in comparisons across laboratories. For norepinephrine, the coefficient of variation averaged 18.4 percent for COMT, 18.9 percent for HPLC and 68.0 percent for fluorimetric assays (Hjemdahl, 1984). The other techniques had variable results, but within a reasonable range of values (mean values of norepinephrine ranged between 1.4−2.72 nM in one sample, and between 2.81−5.66 nM in a second sample; mean values for epinephrine ranged from 0.1−0.45 nM in one sample, and between 2.5−4.39 nM in a second sample). The variability, even excluding studies using the fluorimetric technique, makes comparison of results from different studies problematic. More recently, radioimmunoassays (RIAs) have been used to measure catecholamines both in plasma and urine, and these methods have had good reliability when compared with HPLC methods (Wassell *et al.*, 1999). Reliance on a single technique, or better still, the designation of specific "research-quality" laboratories with strong steps taken for consistent monitoring of inter-laboratory consistency, much like is done for blood lipid analysis through the NIH national laboratory system, would greatly improve comparability of study results.

Another concern for assays is the cost. For HPLC analyses, costs per sample commonly run to $30.00 (US) per sample or higher. Development of a laboratory that has high standards of reliability and validation, but that provides results at low cost would be a major contribution to the ability of field workers to carry out stress measurements utilizing urinary catecholamines.

Relationship between catecholamines and other stress measures

There is a general relationship between the SAMS and the HPA axes of the stress response. As noted above, CRH release by the amygdala stimulates both SNS activity and the HPA axis. The coupling of the SAMS and HPA response is, like the SAMS response itself, not an all-or-nothing phenomenon. The two stress axes are connected through brain mechanisms of activation, with noradrenergic central nervous system (CNS) neurons stimulating the HPA response and CRH secreting CNS neurons stimulating the SAMS response. Epinephrine also stimulates CRH secreting neurons in the hypothalamus (Spinedi *et al.*, 1988). The two stress axes affect each other through interactions in other parts of the body as well. For instance, cortisol, an adrenal corticosteroid hormone released in the HPA response, sensitizes β-adrenergic receptors, with the effect lasting several hours (Davies and Lefkowitz, 1984). Adrenal corticosteroids also up-regulate the enzymes dopamine-β-hydroxylase and phenylethanolamine N-methyltransferase (PNMT) that catalyze epinephrine formation in the adrenal medulla (Axelrod and Reisine, 1984).

The general notion that SAMS activation is located in the adrenal medulla and the HPA response occurs in the adrenal cortex may be too simplistic. While the two parts of the adrenal gland derive from completely different embryological tissues, there are interactions among cells in the medulla and cortex. The catecholamine secreting chromaffin cells are mainly found in the medulla, but are also found in all zones of the adrenal cortex (Ehrhart-Bornstein *et al.*, 1998). Cells characteristic of the cortex are also found in the medulla. This proximity of cells from the two stress axes allows for interaction. Catecholamines have both acute and long-term stimulatory effects on corticosteroid secretion in the adrenal cortex (Güse-Behling *et al.*, 1992; Ehrhart-Bornstein *et al.*, 1998), with the catecholamines stemming both from chromaffin cell secretion and from secretion by nerves within the adrenal cortex. One effect of the catecholamines released by the nerves in the cortex is regulation of diurnal variation in cortisol production (Dijkstra *et al.*, 1996). On the other hand, cortisol secretion within the adrenal medulla stimulates the conversion of dopamine into epinephrine. In sum, the intermingling of adrenal medullary and cortical cells within the adrenal gland leads to mutual interaction between the two stress response axes.

Outside the adrenal gland, corticosteroid hormones potentiate the effects of catecholamines on the cardiovascular system by slowing catecholamine reuptake and enhancing the binding capacity of β-adrenergic receptors (Sapolsky *et al.*, 2000), although these hormones may actually constrain the cardiovascular response to stress, preventing over-reactions (Kvetnansky *et al.*, 1993). However, corticosteroids may also inhibit catecholamine release in certain stress situations (Komesaroff and Funder, 1994).

While the SAMS and HPA axes are closely connected in their actions, responses are not identical. Part of this is due to the time course of response to a stressful stimulus as noted above. Mental challenges and hypoglycemia appear to predominately activate epinephrine and HPA responses, while orthostasis predominately activates norepinephrine secretion, and epinephrine is the main response noted during fainting (Goldstein, 1987). Also, responses of the HPA axis have been shown to habituate much more rapidly to stressful stimuli than does the SAMS both in animal models and in humans (Terrazzino *et al.*, 1995; Schommer *et al.*, 2003), and therefore the two axes diverge in responses to repeated or chronic stressors. The two stress axes also differ in responses to specific stresses and in individuals with certain characteristics. For instance, combat veterans diagnosed with PTSD tend to have high catecholamine and low cortisol levels, with a particularly high norepinephrine/cortisol ratio compared to controls (Mason *et al.*, 1986, 1988; Yehuda *et al.*, 1992).

Individual characteristics that modify catecholamine levels

Demographic characteristics and habitual behaviors have some influence on catecholamine excretion rates independent of stress. These include age, sex, body size and composition, and certain habitual behaviors.

Age

With advancing age, physiological changes occur that can impact catecholamine excretion, such as possibly decreased norepinephrine uptake by neurons, but age changes are more likely due to changes in body composition and physical fitness, and increased prevalence of illness (Ng *et al.*, 1994). Results of some studies suggest that older people tend to excrete greater amounts of catecholamines when exposed to stress (Ziegler *et al.*, 1976; Goldstein *et al.*, 1983), although other studies do not show such differences (Barnes *et al.*, 1982; Ng *et al.*, 1994). It is difficult to disentangle the differing perceptions people may have toward a potential stressful situation based upon experience in a given cohort versus physiological changes due to aging (James and Brown, 1997).

Sex

Women tend to excrete less epinephrine than men when exposed to stress, although there is little or no sex difference in catecholamine excretion levels under control conditions (Collins and Frankenhaeuser, 1978; Frankenhaeuser, 1983; Polefrone and Manuck, 1987). While some sex differences may be due to the effect of steroid sex hormones or other biological differences (Wasilewska *et al.*, 1980; Tersman *et al.*, 1991), most sex differences are apparently due to sociocultural differences associated with gender. In particular, sex differences in catecholamine response under certain stress situations may reflect culturally based variation in attitudes toward these circumstances between genders (James and Brown, 1997). In stressful conditions that are apparently more gender-neutral, such as doing mental arithmetic tests, no gender differences are present (Jones *et al.*, 1996). One study observed a differential sympathetic response to a mental stress test in women when they were in different phases of their menstrual cycle (Hastrup and Light, 1984). There may also be a sex difference in the age-related change in urinary catecholamine excretion (Aslan *et al.*, 1981).

Body size

Body size also has an effect on catecholamine levels. For instance, obese people showed a greater increase in plasma catecholamine levels in response to posture changes and an isometric handgrip exercise than did non-obese controls (Sowers *et al.*, 1982). A mechanism for this may be an interaction between leptin and sympathetic activation, with leptin stimulating SNS activity (Eikelis *et al.*, 2003). Obese people have higher sympathetic nerve firing rates in skeletal muscles and greater renal sympathetic tone than non-obese people (Esler *et al.*, 2003). Formerly obese people who have lost weight have a lowered catecholamine response to fasting than do either never obese or currently obese subjects (Leibel *et al.*, 1991); this reflects their greater caloric efficiency that results from dieting. There is also evidence for a lower level of SNS activity on average among Pima Indians, with this lowered activity associated with weight gain and obesity in this population (Tatarani *et al.*, 1997).

Caffeine, alcohol and nicotine

Habitual behaviors, such as ingestion of caffeine and alcohol, and inhaling of nicotine, affect catecholamine levels. Caffeine ingestion, commonly through drinking coffee or certain soft drinks, elevates plasma epinephrine (Bondi *et al.*, 1999; Kamimori *et al.*, 2000), but has little effect on norepinephrine levels. Heavy drinking of alcohol results in an increase in plasma catecholamines and down-regulation of β-adrenergic receptors (Johnson *et al.*, 1986; Mäki *et al.*, 1998). Similarly, smoking leads to significant increases in plasma epinephrine and norepinephrine levels (Walker *et al.*, 1999).

Because of the various factors that can affect urinary catecholamine levels, field protocols need to be established that account for them. Table 4.3 shows some of the considerations that should be taken in establishing a protocol for the measurement of urinary catecholamines in a field study.

Psychological aspects of SAMS activation

Psychologists have undertaken numerous laboratory-based as well as field-based experiments that have allowed understanding of the factors that induce SAMS activation, and particularly catecholamine elevation in plasma and urine. The earliest work in this area was carried out by Euler and co-workers on military personnel (Euler and Lundberg, 1954), where they found an increase in epinephrine but not in norepinephrine among aircraft passengers, but an increase in both hormones in pilots. Follow-up studies in automobile drivers showed increases in norepinephrine in particular, and it was suggested that norepinephrine elevation is associated with aggressive emotions (Carruthers, 1976).

Studies in people undergoing stress in natural settings have included hospital admission; public speaking; engaging in hazardous situations, including dangerous sports such as parachuting; and the viewing of violent films. All have been associated with increases in epinephrine secretion, and this increase occurs in anticipation of the stressful situation (James *et al.* 1989). Intensity of mood rather than the specific type of mood seems to be of most importance in elevating catecholamines, as

Table 4.3. *Some considerations in establishing research protocols for field studies involving urinary catecholamine measurement*

A. Sample collection

1. Timed samples based on relevant divisions of daily activities; usually of about 4 hours duration or 8 hours for overnight samples;
2. Accurate recording of times, either through careful instructions given to participants or by having trained assistants to collect samples and record the times;
3. Collection over multiple days when possible;
4. Use of appropriate containers: inconspicuous, leak-free, sex-appropriate

---

B. Sample storage

1. Use of appropriate preservatives (sodium metabisulfite or acidification);
2. Saving of multiple aliquots from each well-mixed sample;
3. Freeze samples/aliquots as soon as possible; usually within 24 hours

---

C. Assay technique

1. Radioimmunoassay, radioenzymatic techniques, or high performance liquid chromatography with electrochemical detection;
2. Use of blinded replicates, and use of un-blinded replicates run on separate days;
3. Where possible, sending a selected number of replicates to a reference lab

---

D. Participant compliance

1. Examine outliers in distributions of hormone/volume ratios;
2. Creatinine nomograms (James *et al.*, 1988);
3. Observations by project staff;
4. Participant self-reports

E. Collection of additional information

1. Dietary data (caffeine, alcohol, other drug use, general dietary intake);
2. Demographic information (age, sex);
3. Body size and composition;
4. Smoking;
5. Physical activity;
6. Emotional events and moods;
7. Climactic variables (temperature, humidity)

After: Pearson *et al.*, (1993).

they are also elevated when people view films that elicit pleasant feelings (Froberg *et al.*, 1971).

Some of the major psychological characteristics of stimuli that invoke a stress response, including increase in catecholamines, appear to be threat (Lazarus, 1966; Gross, 1970), novelty (Frankenhaeuser, 1975), unpredictability (Frankenhaeuser and Rissler, 1970; Glass and Singer, 1972), and perceived lack of control (Glass and Singer, 1972; Frankenhaeuser, 1973). Physical factors that consistently elicit catecholamine elevations include noise (Welch and Welch 1969; Babisch *et al.*, 2001; Babisch, 2003), pain (Gliner, 1972), peripheral cold exposure (Victor *et al.*, 1987), and acute exposure to high altitude hypoxia (Hoon *et al.*, 1976).

Occupationally based stress also leads to elevations in catecholamines, as seen in workers who shifted from a salaried to a piecework form of compensation (Levi, 1964) and in machine-paced versus human-paced workers in a sawmill (Frankenhaeuser and Gardell, 1976). There is little direct association between measures of job strain (defined as high job demands with little perceived control) and catecholamines. Commuting to and from work, however, elevates epinephrine levels, but this is related to crowding and other perceived conditions of the commuting (Lundberg, 1976). Bus drivers have elevated catecholamine excretion rates during rush-hour times, as opposed to less traffic-congested times (Evans and Carrere, 1991), with this thought to be related to perceived control over the situation. Much of the response to psychological stress is mediated by situational attributes, and for most people, most of the time, these mediators are sociocultural in nature.

Anthropological field studies of catecholamine responses to stress

There have not been many studies utilizing catecholamine excretion as a measure of stress in anthropological field studies. Here, a few examples will be chosen to indicate the type of studies that have availed themselves of this stress measure.

Among the first anthropological field studies utilizing catecholamine excretion rates was a study in Oxfordshire, United Kingdom, by G.A. Harrison and associates that began in the 1970s. This research focused on the sociocultural characteristics that were associated with elevations in catecholamine excretion rates (Harrison, 1995). Timed urine samples were collected over three periods (morning, midday and early evening) on a work day and a non-work day, and the samples were analyzed for catecholamine excretion rates. Catecholamine excretion was seen to vary by occupation, with managers and professionals having the highest rates, and also to change according to whether collection took place on a work day or holiday (Jenner *et al.*, 1979, 1980). Smoking and drinking caffeine was also associated with elevated catecholamine levels (Reynolds *et al.*, 1981; Harrison, 1995). Levels were lower in the morning than at other times. Epinephrine excretion rates were positively associated with worrying and frustration, and negatively associated with reported life satisfaction in women (Harrison *et al.*, 1981). Among men, occupationally related stress and mental tiredness were associated with elevated epinephrine, while physical tiredness and boredom were related to lowered epinephrine excretion rates (Reynolds *et al.*, 1981).

Several studies have examined the effect of culture change, whether *in situ* or due to migration, on catecholamine excretion rates. For instance, Brown (1981, 1982) collected 24-hour urine samples over a three-day period from immigrant Filipino-Americans residing in a community near Honolulu and discovered that catecholamine excretion rates were elevated in immigrants with an intermediate degree of Americanization compared with those with either very little or a great deal of contact with American culture. In a study of Filipino-American immigrant nurses in the Hilo region of Hawaii, it was noted that nurses who had resided in the USA for a long time period had higher norepinephrine excretion rates than those who were more recent immigrants (Brown and James, 2000).

Gary James (James *et al.*, 1985, 1987) conducted a study on the effect of modernization on Samoans living in Apia, collecting timed urine samples at mid-morning and overnight from men who were either from a rural village or from one of three urban groups: manual laborers, sedentary workers, and college students. He noted that villagers tended to have relatively low catecholamine excretion rates at both times, while urban laborers had relatively low rates for the overnight sample, but higher rates for the mid-morning sample. The other two urban groups had relatively high catecholamine excretion rates at both times. James suggested that the villagers were less modernized, and the urban laborers were exposed to modern lifestyles mostly during the work day, retaining traditional lives in the evening. Thus, high catecholamines were associated with greater modernization. In another study of stress and modernization among Samoans it was observed that lower rates of epinephrine excretion were found in people with more traditional lifestyles (Martz *et al.*, 1984; Hanna *et al.*, 1986). Jenner *et al.* (1987) compared catecholamine excretion rates from several populations and noted that values from "traditional" groups were lower than those from more modernized groups. Thus, transition into a more modernized lifestyle appears to be associated with an increase in urinary catecholamines, whether that modernization occurs due to migration or to changes in one's circumstances without need for relocation.

Other stress studies among Samoans living in what was then Western Samoa and Hawaii showed that relatively high epinephrine excretion rates were found in women who spent more time in social interactions, while just the reverse was found for women living in American Samoa where women with less time in social interactions had relatively elevated epinephrine levels (Pearson *et al.*, 1990, 1993).

One final example of an anthropological field study of stress and catecholamine levels was done among three communities of aboriginal Australians. It was noted that in the community where alcohol was "freely available," epinephrine excretion rates were higher than in the other communities (Schmitt *et al.*, 1995). Within one community, urine collections were made during four days of a single week, with collections occurring over a timed two hour period between 3 and 5 p.m. Epinephrine excretion rates were significantly higher on two days of the week, Thursday and Friday, compared to the two other days, Monday and Tuesday, during which catecholamine levels were monitored (Schmitt *et al.*, 1998). The days in which epinephrine levels were elevated were associated with arrival of pay checks, which was followed by intense gambling activity which continued for about two days, after which such activity declined considerably. It is thought that the intense emotions involved in the gambling were a causal factor in the elevated epinephrine excretion rate.

Conclusion

Although anthropological field studies of catecholamine excretion have been carried out for nearly thirty years, the studies are relatively few, with most centered in the Pacific region. This may be due to the expense of analyses when relying upon commercial laboratories to carry out analyses, or the need for complex equipment such as HPLCs which are unfamiliar to most anthropologists. Whatever the cause, the paucity of studies is unfortunate, as analysis of urinary catecholamines fairly accurately reflects sympathetic system activity, is non-invasive, and can be carried out under conditions where refrigeration is not available for 24 hours or more. Catecholamines can be used under many circumstances to serve as a measure of acute stress both for individuals and groups. Studies involving catecholamines have potential for revealing the psychological import of specific daily events, particularly those lasting for an hour or more, and are free of the bias due to cultural differences in what it is believed to be proper to report in responses to questionnaires and many psychological instruments (Brown, 1981). It is clear that urinary catecholamines represent a reasonable approach to study of

SAMS activity for people who are being monitored while doing normal daily activities. Other measures such as CSF or plasma catecholamines, while reflecting SAMS activity in a fairly direct manner, have serious drawbacks due to the invasive procedures necessary for their collection. These procedures actually change SAMS activity, and thus eliminate the ability to observe the relationship between normal activities and the stress response.

References

Akerstedt, T., Gillberg, M., Hjemdahl, P. *et al.* (1983). Comparisons of urinary and plasma catecholamine responses to mental stress. *Acta Physiologica Scandinavica*, **117**, 19–26.

Allen, J. P. and Allen, C. F. (1974). Role of the amygdaloid complexes in the stress-induced release of ACTH in the rat. *Neuroendocrinology*, **15**, 220–30.

Aslan, S., Nelson, L., Carruthers, M. and Lader, M. (1981). Stress and age effects on catecholamines in normal subjects. *Journal of Psychosomatic Research*, **25**, 33–41.

Axelrod, J. and Reisine, T. D. (1984). Stress hormones: their interaction and regulation. *Science*, **224**, 452–9.

Babisch, W. (2003). Stress hormones in the research on cardiovascular effects of noise. *Noise Health*, **5**, 1–11.

Babisch, W., Fromme, H., Beyer, A. and Ising, H. (2001). Increased catecholamine levels in urine in subjects exposed to road traffic noise: the role of stress hormones in noise research. *Environmental International*, **26**, 475–81.

Barnes, R. F., Raskind, M., Gumbrecht, G. and Halter, J. B. (1982). The effects of age on the plasma catecholamine response to mental stress in man. *Journal of Clinical and Endocrinology and Metabolism*, **54**, 64–9.

Best, J. D. and Halter, J. B. (1982). Release and clearance rates of epinephrine in man: importance of arterial measurements. *Journal of Clinical Endocrinological Metabolism*, **55**, 263–8.

Bondi, M., Grugni, G., Velardo, A. *et al.* (1999). Adrenomedullary response to caffeine in prepubertal and pubertal obese subjects. *International Journal of Obesity and Related Metabolic Disorders*, **23**, 992–6.

Boomsma, F., Alberts, G., van Eijk, L., Man in't Veld, A. J. and Schalekamp, M. A. (1993). Optimal collection and storage conditions for catecholamine measurements in human plasma and urine. *Clinical Chemistry*, **39**, 2503–8.

Brown, D. E. (1981). General stress in anthropological fieldwork. *American Anthropologist*, **83**, 74–92.

(1982). Physiological stress and culture change in a group of Filipino-Americans: a preliminary investigation. *Annals of Human Biology*, **9**, 553–63.

Brown, D. E. and James, G. D. (2000). Physiological stress responses in Filipino-American immigrant nurses: the effects of residence time, life-style, and job strain. *Psychosomatic Medicine*, **62**, 394–400.

Calaresu, F. R. and Yardley, C. P. (1988). Medullary basal sympathetic tone. *Annual Review of Physiology*, **50**, 511–24.

Callingham, B. A. (1975). Catecholamines in blood. In *Handbook of Physiology: Section 7. Endocrinology. Volume VI. Adrenal Gland*, ed. R. O. Greep and E. B. Astwood. Washington, DC: American Physiological Society, pp. 427–45.

Calogero, A. E., Galluci, W. T., Chrousos, G. P. and Gold, P. W. (1988). Catecholamine effects upon rat hypothalamic corticotropin-releasing hormone secretion in vitro. *Journal of Clinical Investigation*, **82**, 839–46.

Cannon, W. B. (1915) *Bodily Changes in Pain, Hunger, Fear and Rage: An Account of Recent Researches into the Functions of Emotional Excitement*. New York: Appleton.

Carruthers, M. (1976). Biochemical responses to environmental stress. In *Man in Urban Environments*, ed. G. A. Harrison and J. B. Gibson. Oxford: Oxford University Press, pp. 247–73.

Carruthers, M., Taggert, P., Conway N., Bates, D. and Somerville, W. (1970). Validity of plasma catecholamine estimation. *Lancet*, **2**, 62–7.

Charmandari, E., Kino, T., Souvatzoglou, E. and Chrousos, G. P. (2003). Pediatric stress: hormonal mediators and human development. *Hormone Research*, **59**, 161–79.

Chrousos, G. P. and Gold, P. W. (1992). The concepts of stress and stress system disorders: overview of physical and behavioral homeostasis. *Journal of the American Medical Association*, **267**, 1244–52.

Collins, A. and Frankenhaeuser, M. (1978). Stress responses in male and female engineering students. *Journal of Human Stress*, **4**, 43–8.

Davidson, L., Vandongen, R. and Beilin, L. J. (1981). Effect of eating bananas on plasma free and sulfate-conjugated catecholamines. *Life Science*, **29**, 1773–8.

Davies, A. and Lefkowitz, R. (1984). Regulation of beta-adrenergic receptors by steroid hormones. *Annual Review of Physiology*, **46**, 119–30.

Dijkstra, I., Binnekade, R. and Tilders, F. J. H. (1996). Diurnal variation in resting levels of corticosterone is not mediated by variation in adrenal responsiveness to adrenocorticotrophin but involves splanchnic nerve integrity. *Endocrinology*, **137**, 540–7.

Dimsdale, J. E. and Ziegler, M. G. (1991). What do plasma and urinary measures of catecholamines tell us about human response to stressors? *Circulation*, **83** (Supplement II), II-36–42.

Dunne, J. W., Davidson, L., Vandongen, R., Beilin, L. J. and Rogers, P. (1983). The effect of ascorbic acid on plasma sulfate conjugated catecholamines after eating bananas. *Life Science*, **33**, 1511–17.

Ehrhart-Bornstein, M., Hinson, J. P., Bornstein, S. R., Scherbaum, W. A. and Vinson, G. P. (1998). Intraadrenal interactions in the regulation of adrenocortical steroidogenesis. *Endocrine Reviews*, **19**, 101–43.

Eide, I., Kolloch, R., De Quattro, V. *et al.* (1979). Raised cerebrospinal fluid norepinephrine in some patients with primary hypertension. *Hypertension*, **1**, 255–60.

Eikelis, N., Schlaich, M., Aggarwal, A., Kaye, D. and Esler, M. (2003). Interactions between leptin and the human sympathetic nervous system. *Hypertension*, **41**, 1072–9.

Esler, M. (1993). Clinical application of noradrenaline spillover methodology: delineation of regional human sympathetic nervous responses. *Pharmacological Toxicology*, **73**, 243–53.

(1995). The sympathetic nervous system and catecholamine release and plasma clearance in normal blood pressure control, in aging, and in hypertension. In *Hypertension: Pathophysiology, Diagnosis, and Management*, 2nd edn, ed. J. H. Laragh and B. M. Brenner. New York: Raven Press, pp. 755–73.

Esler, M., Jennings, G., Korner, P. *et al.* (1988). Assessment of human sympathetic nervous system activity from measurements of norepinephrine turnover. *Hypertension*, **11**, 3–20.

Esler, M., Jennings, G., Lambert, G. et al. (1990). Overflow of catecholamine neurotransmitters to the circulation: source, fate and functions. *Physiological Reviews*, **70**, 963–85.

Esler, M., Lambert, G., Brunner-LaRocca, H. P., Vaddadi, G. and Kaye, D. (2003). Sympathetic nerve activity and neurotransmitter release in humans: translation from pathophysiology into clinical practice. *Acta Physiologica Scandinavica*, **177**, 275–84.

Euler, U. S. von (1967). Adrenal medullary secretion and its neural control. In *Neuroendocrinology*, ed. L. Martini and W. F. Ganong. New York: Academic Press, pp. 283–333.

Euler, U. S. von and Lundberg, U. (1954). Effect of flying on the epinephrine excretion in air force personnel. *Journal of Applied Physiology*, **6**, 551–5.

Evans, G. W. and Carrere, S. (1991). Traffic congestion, perceived control, and psychophysiological stress among urban bus drivers. *Journal of Applied Psychology*, **76**, 658–63.

Fitzgerald, G. A., Hossman, V., Hamilton, C. A. *et al.* (1979). Interindividual variation in kinetics of infused epinephrine. *Clinical Pharmacology & Therapeutics*, **26**, 669–75.

Foote, S. L., Bloom, F. E. and Aston-Jones, G. (1983). Nucleus locus coeruleus: new evidence of anatomical and physiological specificity. *Physiological Reviews*, **63**, 844–914.

Frankenhaeuser, M. (1973). Experimental approaches to the study of human behavior as related to neuroendocrine functions. In *Society, Stress and Disease*, ed. L. Levi. New York: Oxford University Press, pp. 22–35.

(1975). Sympathetic-adrenomedullary activity, behavior and the psychosocial environment. In *Research in Psychophysiology*, ed. P. H. Venables and M. J. Christie. New York: John Wiley, pp. 71–94.

(1983). The sympathetic-adrenal and pituitary-adrenal response to challenge, comparisons between the sexes. In *Biobehavioral Bases of Coronary Heart*

Disease, ed. T. M. Dembrowski, T. H. Schmidt and G. Blumchen. Basel: Karger, pp. 91–105.

Frankenhaeuser, M. and Gardell, B. (1976). Underload and overload in working life. A multidisciplinary approach. *Journal of Human Stress*, **2**, 35–46.

Frankenhaeuser, M. and Rissler, A. (1970) Effects of punishment on catecholamine release and efficiency of performance. *Psychopharmacologia*, **17**, 378–90.

Frishman, W. H. and Lazar, E. J. (1990). Reduction of mortality, sudden death and non-fatal reinfarction with β-adrenergic blockers in survivors of acute myocardial infarction: a new hypothesis regarding the cardioprotective action of β-adrenergic blockade. *American Journal of Cardiology*, **66**, 66–70G.

Froberg, J., Karlsson, C., Levi, L. and Lidberg, L. (1971). Physiological and biochemical stress reactions induced by psychosocial stimuli. In *Society, Stress and Disease. Volume I: The Psychosocial Environment and Psychosomatic Diseases*, ed. L. Levi. London: Oxford University Press, pp. 280–95.

Glass, D. C. and Singer, J. E. (1972). *Urban Stress*. New York: Academic Press.

Gliner, J. A. (1972). Predictable vs. unpredictable shock: preference behavior and stomach ulceration. *Physiology & Behavior*, **9**, 693–8.

Goldstein, D. S. (1987). Stress-induced activation of the sympathetic nervous system. *Baillière's Clinical Endocrinology and Metabolism*, **1**, 253–78.

Goldstein, D. S., Lake, C. R., Chernow, B. *et al.* (1983). Age-dependence of hypertensive-normotensive differences in plasma norepinephrine. *Hypertension*, **5**, 100–104.

Gross, E. (1970) Work, organization, and stress. In *Social Stress*, ed. S. Levine and N. Scotch. Chicago: Aldine Publishing Company, pp. 54–110.

Güse-Behling, H., Ehrhart-Bornstein, M., Bornstein, S. R. *et al.* (1992). Regulation of adrenal steroidogenesis by adrenaline: expression of cytochrome P450 genes. *Journal of Endocrinology*, **135**, 229–37.

Guyton, A. C. (1971). *Textbook of Medical Physiology*, 4th edn. Philadelphia: W. B. Saunders Company.

Hague, C., Chen, Z., Uberti, M. and Minneman, K. P. (2003). Alpha-1 adrenergic receptor subtypes: non-identical triplets with differing dancing partners? *Life Sciences*, **74**, 411–18.

Hanna, J. M., James, G. D. and Martz, J. (1986). Hormonal measures of stress. In *The Changing Samoans: Behavior and Health in Transition*, ed. P. T. Baker, J. M. Hanna and T. S. Baker. Oxford: Oxford University Press, pp. 203–21.

Harrison, G. A. (1995). *The Human Biology of the English Village*. Oxford: Oxford University Press.

Harrison, G. A., Palmer, C. D., Jenner, D. A. and Reynolds, V. (1981). Association between rates of urinary catecholamine excretion and aspects of lifestyle among adult women in some Oxfordshire villages. *Human Biology*, **53**, 617–33.

Harrison, T., Samuel, B. U., Akompong, T. *et al.* (2003). Erythrocyte G protein-coupled receptor signaling in malarial infection. *Science*, **301**, 1734–6.

Hastrup, J. L. and Light, K. C. (1984). Sex differences in cardiovascular stress responses: modulation as a function of menstrual cycle phases. *Journal of Psychosomatic Research*, **28**, 475–83.

Hjemdahl, P. (1984). Inter-laboratory comparison of plasma catecholamine determinations using several different assays. *Acta Physiologica Scandanavica*, **527** (Supplement), 43–54.

Hoon, R. S., Sharma, S. C., Balasubramanian, V., Chadha, K. S. and Mathew O. P. (1976). Urinary catecholamine excretion on acute exposure to high altitude (3,658 m). *Journal of Applied Physiology*, **41**, 631–3.

Insel, P. A. (1996). Adrenergic receptors – evolving concepts and clinical implications. *New England Journal of Medicine*, **334**, 580–5.

James, G. D. and Brown, D. E. (1997). The biological stress response and lifestyle: catecholamines and blood pressure. *Annual Review of Anthropology*, **26**, 313–35.

James, G. D., Baker, P. T., Jenner, D. A. and Harrison, G. A. (1987). Variation in lifestyle characteristics and catecholamine excretion rates among young Western Samoan men. *Social Science and Medicine*, **25**, 981–6.

James, G. D., Crews, D. E. and Pearson, J. (1989). Catecholamines and stress. In *Human Population Biology: A Transdisciplinary Science*, ed. M. A. Little and J. D. Haas. Oxford: Oxford University Press, pp. 280–95.

James, G. D., Jenner, D. A., Harrison, G. A. and Baker, P. T. (1985). Differences in catecholamine excretion rates, blood pressure and lifestyle among young Western Samoan men. *Human Biology*, **57**, 635–47.

James, G. D., Schlussel, Y. R. and Pickering, T. G. (1993). The association between daily blood pressure and catecholamine variability in normotensive working women. *Psychosomatic Medicine*, **55**, 55–60.

James, G. D., Sealey, J. E., Alderman, M. *et al.* (1988). A longitudinal study of urinary creatinine and creatinine clearance in normal subjects. *American Journal of Hypertension*, **1**, 124–31.

Januszewicz, W., Sznajderman, M., Wocial, B., Feltynowski, T. and Klonowicz, T. (1979). The effect of mental stress on catecholamines, their metabolites and plasma renin activity in patients with essential hypertension and in healthy subjects. *Clinical Science*, **57**, 229S–31S.

Jenner, D. A., Harrison, G. A., Prior, I. A. M. *et al.* (1987). Inter-population comparisons of catecholamine excretion. *Annals of Human Biology*, **14**, 1–9.

Jenner, D. A., Reynolds, V. and Harrison, G. A. (1979). Population field studies of catecholamines. In *Response to Stress: Occupational Aspects*, ed. C. MacKay and T. Cox. London: IPC Science and Technology Press, pp. 112–19.

(1980). Catecholamine excretion rates and occupation. *Ergonomics*, **23**, 237–46.

Johnson, R. H., Eisenhofer, G. and Lambie, D. G. (1986). The effects of acute and chronic ingestion of ethanol on the autonomic nervous system. *Drug and Alcohol Dependence*, **18**, 319–28.

Jones, P. P., Spraul, M., Matt, K. S. *et al.* (1996). Gender does not influence sympathetic neural reactivity to stress in healthy humans. *American Journal of Physiology*, **270**, H350–7.

Kamimori, G. H., Penetar, D. M., Headley, D. B. *et al.* (2000) Effect of three caffeine doses on plasma catecholamines and alertness during prolonged wakefulness. *European Journal of Clinical Pharmacology*, **56**, 537–44.

Kim, J. J., Lee, H. J., Han, J.-S. and Packard, M. G. (2001). Amygdala is critical for stress-induced modulation of hippocampal long-term potentiation and learning. *Journal of Neuroscience*, **21**, 5222–8.

Komesaroff, P. and Funder, J. (1994). Differential glucocorticoid effects on catecholamine responses to stress. *American Journal of Physiology*, **266**, E118–23.

Kvetnansky, R., Fukuhara, K., Pacak, K. *et al.* (1993). Endogenous glucocorticoids restrain catecholamine synthesis and release at rest and during immobilization stress in rats. *Endocrinology*, **133**, 1411–19.

Lafantan, M. and Berlan, M. (1993). Fat cell adrenergic receptors and the control of white and brown fat cell function. *Journal of Lipid Research*, **34**, 1057–91.

Lazarus, R. S. (1966). *Psychological Stress and the Coping Process*. New York: McGraw-Hill.

Leibel, R. L., Berry, E. M. and Hirsch, J. (1991). Metabolic and hemodynamic responses to endogenous and exogenous catecholamines in formerly obese subjects. *American Journal of Physiology*, **260**, R785–91.

Levi, L. (1964) The stress of everyday work as reflected in productiveness, subjective feelings, and urinary output of adrenaline and noradrenaline under salaried and piece-work conditions. *Journal of Psychosomatic Research*, **8**, 199–202.

Lewis, G. P. (1975). Physiological mechanisms controlling secretory activity of adrenal medulla. In *Handbook of Physiology: Section 7. Endocrinology. Volume VI. Adrenal Gland*, ed. R. O. Greep and E. B. Astwood. Washington, DC: American Physiological Society, pp. 309–489.

Li, H.-Y., Ericsson A. and Sawchenko, P. E. (1996). Distinct mechanisms underlie activation of hypothalamic neurosecretory neurons and their medullary catecholaminergic afferents in categorically different stress paradigms. *Proceedings of the National Academy of Sciences*, **93**, 2359–64.

Link, R. E., Desai, K., Hein, L. *et al.* (1996). Cardiovascular regulation in mice lacking α_2-adrenergic receptor subtypes b and c. *Science*, **273**, 803–5.

Lundberg, U. (1976). Urban commuting: crowdedness and catecholamine excretion. *Journal of Human Stress*, **2**, 26–32.

MacMillan, L. B., Hein, L., Smith, M. S., Piascik, M. T. and Limbird, L. E. (1996). Central hypotensive effects of the α_{2A}-adrenergic receptor subtype. *Science*, **273**, 801–3.

Mäki, T., Toivonen, L., Koskinen, P. *et al.* (1998). Effect of ethanol drinking, hangover, and exercise on adrenergic activity and heart rate variability in patients with a history of alcohol-induced atrial fibrillation. *American Journal of Cardiology*, **82**, 317–22.

Martz, J., Hanna, J. M. and Howard, S. A. (1984). Stress in daily life. Evidence from Samoa. *American Journal of Physical Anthropology*, **63**, 191–2.

Mason, J. W., Giller, E. L., Kosten, T. R. and Harkness, L. (1988). Elevation of urinary norepinephrine/cortisol ratio in posttraumatic stress disorder. *Journal of Nervous and Mental Disease*, **176**, 498–502.

Mason, J. W., Giller, E. L., Kosten, T. R., Ostroff, R. B. and Podd, L. (1986) Urinary free cortisol levels in post-traumatic stress disorder patients. *Journal of Nervous and Mental Disease*, **174**, 145–9.

McEwen, B. S. (1998). Stress, adaptation and disease. *Annals of the New York Academy of Science*, **840**, 33–44.

(2003). Interacting mediators of allostasis and allostatic load: towards an understanding of resilience in aging. *Metabolism*, **52**(Supplement 2), 10–16.

McEwen, B. S. and Stellar, E. (1993). Stress and the individual: mechanisms leading to disease. *Archives of Internal Medicine*, **153**, 2093–101.

Miki, K. and Sudo, A. (1998). Effect of urine pH, storage time, and temperature on stability of catecholamines, cortisol, and creatinine. *Clinical Chemistry*, **44**, 1759–62.

Nantel, F., Marullo, S., Krief, S., Strosberg, A. D. and Bouvier, M. (1994). Cell-specific down-regulation of the β3-adrenergic receptor. *Journal of Biological Chemistry*, **269**, 13148–55.

Ng, A. V., Callister, R., Johnson, D. G. and Seals, D. R. (1994). Sympathetic neural reactivity to stress does not increase with age in healthy humans. *American Journal of Physiology*, **267**, H344–53.

Ninomiya, I., Irasawa, A. and Nisimaru, N. (1973). Nonuniformity of sympathetic nerve activity to the skin and kidney. *American Journal of Physiology*, **224**, 256–64.

Pearson, J. D., Hanna, J. M., Fitzgerald, M. H. and Baker, P. T. (1990). Modernization and catecholamine excretion of young Samoan adults. *Social Science and Medicine*, **31**, 729–36.

Pearson, J. D., James, G. D. and Brown, D. E. (1993). Stress and changing lifestyles in the Pacific: physiological stress responses of Samoans in urban and rural settings. *American Journal of Human Biology*, **5**, 49–60.

Pederson, B. K. and Hoffman-Goetz, L. (2000). Exercise and the immune system: regulation, integration, and adaptation. *Physiological Reviews*, **80**, 1055–81.

Piascik, M. T. and Perez, D. M. (2001). α_1-adrenergic receptors: new insights and directions. *Journal of Pharmacology and Experimental Therapeutics*, **298**, 403–10.

Polefrone, J. M. and Manuck, S. B. (1987). Gender differences in cardiovascular and neuroendocrine response to stressors. In *Gender and Stress*, ed. R. C. Barnett, L. Biener and G. K. Barach. New York: Free Press, pp. 13–38.

Psaty, B. M., Smith, N. L., Siscovick, D. S. *et al.* (1997). Health outcomes associated with antihypertensive therapies used as first-line agents. A systematic review and meta-analysis. *Journal of the American Medical Association*, **277**, 739–45.

Reynolds, V., Jenner, D. A., Palmer, C. D. and Harrison, G. A. (1981). Catecholamine excretion rates in relation to life-styles in the male population of Oxmoor, Oxfordshire. *Annals of Human Biology*, **8**, 197–209.

Sapolsky, R. M. (1994). *Why Zebras Don't Get Ulcers: A Guide to Stress, Stress-Related Diseases, and Coping*. New York: W. H. Freeman and Company.

Sapolsky, R. M., Romero, M. and Munck, A. U. (2000). How do glucocorticoids influence stress responses? Integrating permissive, suppressive, stimulatory, and preparative actions. *Endocrine Reviews*, **21**, 55−89.

Sawchenko, P. E., Li, H.-Y. and Ericsson, A. (2000). Circuits and mechanisms governing hypothalamic responses to stress: a tale of two paradigms. *Progress in Brain Research*, **122**, 61−78.

Schmitt, L. H., Harrison, G. A. and Spargo, R. M. (1998). Variation in epinephrine and cortisol excretion rates associated with behavior in an Australian aboriginal community. *American Journal of Physical Anthropology*, **106**, 249−53.

Schmitt, L. H., Harrison, G. A., Spargo, R. M., Pollard, T. and Ungpakorn, G. (1995). Patterns of cortisol and adrenaline variation in Australian aboriginal communities of the Kimberley region. *Journal of Biosocial Science*, **27**, 107−16.

Schommer, N. C., Hellhammer, D. H. and Kirschbaum, C. (2003). Dissociation between reactivity of the hypothalamus-pituitary-adrenal axis and the sympathetic-adrenal medullary system to repeated psychosocial stress. *Psychosomatic Medicine*, **65**, 450−60.

Silverberg, A. B., Shah, S. D., Haymond, M. W. and Cryer, P. E. (1978). Norepinephrine: hormone and neurotransmitter in man. *American Journal of Physiology*, **234**, E252−6.

Sowers, J. R., Whitfield, L. A., Catania, R. A. *et al.* (1982). Role of the sympathetic nervous system in blood pressure maintenance in obesity. *Journal of Clinical Endocrinology and Metabolism*, **54**, 1181−6.

Spinedi, E., Johnston, C. A., Chisari, A. and Negro-Vilar, A. (1988). Role of central epinephrine on the regulation of corticotrophin releasing factor and adrenocorticotrophin secretion. *Endocrinology*, **122**, 1977−83.

Tatarani, P. A., Young, J. B., Bogardus, C. and Ravussin, E. (1997). A low sympathoadrenal activity is associated with body weight gain and development of central adiposity in Pima Indian men. *Obesity Research*, **5**, 341−7.

Terrazzino, S., Perego, C. and De Simoni, M. G. (1995). Effect of development of habituation to restraint stress on hypothalamic noradrenaline release and adrenocorticotrophin secretion. *Journal of Neurochemistry*, **65**, 263−7.

Tersman, Z., Collins, A. and Eneroth, P. (1991). Cardiovascular responses to psychological and physiological stressors during the menstrual cycle. *Psychosomatic Medicine*, **53**, 185−7.

Victor, R. G., Leimbach, W. N., Jr., Seals, D. R., Wallin, B. G. and Mark, A. L. (1987). Effects of the cold pressor test on muscle sympathetic nerve activity in humans. *Hypertension*, **9**, 429−36.

Walker, J. F., Collins, L. C., Rowell, P. P. *et al.* (1999). The effect of smoking on energy expenditure and plasma catecholamine and nicotine levels during light physical activity. *Nicotine Tobacco Research*, **1**, 365−70.

Wallin, B. G., Sundlöv, G., Eriksson, B.-M. *et al.* (1981). Plasma noradrenaline correlates to sympathetic muscle nerve activity in normotensive man. *Acta Physiologica Scandinavica*, **111**, 69−73.

Ward, M. and Mefford, I. N. (1985). Methodology of studying the catecholamine response to stress. In *Clinical and Methodological Issues in Cardiovascular Psychophysiology*, ed. A. Steptoe, H. Ruddel and H. Neus. Berlin: Springer-Verlag, pp. 131–43.

Wasilewska, E., Kobus, E. and Bargiel, Z. (1980). Urinary catecholamine excretion and plasma dopamine-beta-hydroxylase activity in mental work performed in two periods of menstrual cycle in women. In *Catecholamines and Stress: Recent Advances*, ed. E. Usdin, R. Kvetnansky and I. J. Kopin. New York: Elsevier, pp. 549–54.

Wassell, J., Reed, P., Kane, J. and Weinkove, C. (1999). Freedom from drug interference in new immunoassays for urinary catecholamines and metanephrines. *Clinical Chemistry*, **45**, 2216–23.

Welch, B. L. and Welch, A. S. (eds.) (1969). *Physiological Effects of Noise*. New York: Plenum Press.

Wilkinson, D. J., Thompson, J. M., Lambert, G. W. *et al.* (1998). Sympathetic activity in patients with panic disorder at rest, under laboratory mental stress, and during panic attacks. *Archives General Psychiatry*, **55**, 511–20.

Yehuda, R., Southwick, S., Giller, E. L., Ma, X. and Mason, J. W. (1992). Urinary catecholamine excretion and severity of PTSD symptoms in Vietnam combat veterans. *Journal of Nervous and Mental Disease*, **180**, 321–5.

Ziegler, M., Kennedy, B. and Elayan, H. (1989). Rat renal epinephrine synthesis. *Journal of Clinical Investigation*, **84**, 1130–3.

Ziegler, M. G., Lake, C. R. and Kopin, I. J. (1976). Plasma noradrenaline increases with age. *Nature*, **261**, 333–5.

5 Measuring hormonal variation in the hypothalamic pituitary adrenal axis: cortisol

TESSA M. POLLARD AND GILLIAN H. ICE

What is cortisol?

Cortisol is the main corticosteroid hormone in humans. Its secretion is controlled by the hypothalamic pituitary adrenal (HPA) endocrine axis (Figure 5.1). This pathway is initiated by the release of cortico-trophin releasing hormone (CRH) from the paraventricular nucleus of the hypothalamus. CRH prompts the pituitary to secrete adrenocortico-trophic hormone (ACTH) and ACTH then stimulates the adrenal cortex to produce cortisol. Secretion of cortisol occurs in pulses, the number of which determine cortisol levels in the blood. There is a tight coupling between these last two components of the HPA axis, such that a rise in ACTH strongly predicts a rise in cortisol (Dickerson and Kemeny, 2004). Several feedback loops (Figure 5.1) regulate the activity of the HPA axis, providing sensitive mechanisms for the adjustment of the circulating cortisol level during everyday life (Brunner and Marmot, 1999). For example, cortisol exerts negative feedback effects on the hypothalamus, inhibiting the production of CRF.

The pathway acts more slowly than the SAM pathway which con-trols the release of the other main stress hormones, the catecholamines. In laboratories ACTH levels start to rise within five minutes of expo-sure to a stressor and peak 11–20 minutes from the time of stressor onset (Kirschbaum and Hellhammer, 1989; Dickerson and Kemeny, 2004). Cortisol levels in plasma peak 20–40 minutes after exposure (Kirschbaum and Hellhammer, 1989; Dickerson and Kemeny, 2004). Most of the physiological actions of stress-induced rises in cortisol are

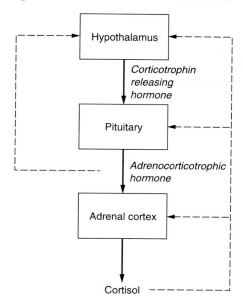

Figure 5.1.   The hypothalamic pituitary adrenal axis — a simplified version. Feedback loops are indicated by dotted lines.

exerted about an hour after the stimulation of the HPA axis. Its effects are mainly mediated via genomic mechanisms, which take some time to come into effect (Sapolsky *et al.*, 2000). Cortisol crosses the cell membrane and then binds to glucocorticoid receptors and mineralo-corticoid receptors in the cytoplasm. The receptor complex then enters the cell nucleus, and causes the activation or repression of particular genes (Munck, 2000). The half-life of cortisol is about 70 minutes (Baum and Grunberg, 1995).

Ultimately the HPA axis is controlled by events initiated by the cerebral cortex. Here environmental and endogenous stimuli act in conjunction with evaluative processes to initiate an emotional response (Lovallo and Thomas, 2000). Emotions can be considered to mediate the effects of appraisal of environmental circumstances and coping resources on autonomic and endocrine systems such as the HPA axis.

What does cortisol do?

Cortisol is essential for the maintenance of normal functions within the body, including growth and development, as well as for diurnal variations in metabolism (Lovallo and Thomas 2000). Cortisol plays a key role in the metabolism of proteins and fat and in gluconeogenesis. In addition, it supports vascular responsiveness, skeletal turnover, muscle function, immune response and renal function (Berne and Levy, 1993). Here we describe the effects of cortisol in relation to stress.

Short-term physiologic effects

Cortisol plays several different types of roles in relation to the body's response to a stressor (Sapolsky *et al.*, 2000). Background levels of cortisol permit the actions of other stress-related systems in the early stages of the stress response, while stress-induced rises in cortisol may stimulate the effects of other systems in some ways. However, an important role of cortisol elevation in response to stress is to rein in the activities of other systems after an appropriate time. Sapolsky *et al.* (2000) also suggest that cortisol elevation helps the body prepare for subsequent stressors.

For example, glucocorticoids augment the actions of catecholamines and other vasoconstrictors on the cardiovascular system and with respect to the mobilization of glucose and lipids. While epinephrine acts quickly to elevate blood glucose levels, cortisol acts more slowly and maintains the high levels of glucose, partly by increasing insulin resistance (Munck, 2000). Cortisol has anti-inflammatory effects, for example, inhibiting the synthesis, release and/or efficacy of cytokines and other mediators that promote immune and inflammatory reactions (Heim *et al.*, 2000; Sapolsky *et al.*, 2000). On the other hand, it has become increasingly clear that cortisol enhances some aspects of immune function. Cortisol appears to act to decrease circulating blood leukocytes while increasing leukocyte numbers at sites of injury or other areas of need, a phenomenon referred to as "stress-induced trafficking" (Bilbo *et al.*, 2002; McEwen, 2002). It is thought to inhibit reproductive function, perhaps by inhibiting gonadotropin-releasing hormone release from the hypothalamus, and the HPA axis may also have direct effects on ovarian function (Nepomnaschy *et al.*, 2005). Cortisol also has effects in the brain, including the promotion of memory formation (McEwen and Wingfield, 2003). This is thought to be adaptive because it may help individuals remember situations which have proved dangerous or threatening.

Links with disease

People with high levels of cortisol associated with Cushing's syndrome or clinical depression show a number of adverse effects, including atrophy of the hippocampus and associated malfunction in memory formation, abdominal obesity, and loss of bone density (Miller and O'Callaghan, 2002; Brown *et al.*, 2004). Such evidence is suggestive about the effects of elevated cortisol levels in individuals without these pathologies.

As detailed in Chapter 1, McEwen and Wingfield (2003) have suggested that repeated stimulation of the cortisol response may lead to allostatic load, which is characterized by chronically elevated cortisol levels as well as changes in other physiological systems. They suggest that in combination with the excess energy consumption characteristic of humans in affluent societies, this can lead to a number of potentially pathological effects. For example, chronically elevated glucocorticoids are expected to impede the promotion of glucose uptake by insulin. In response to this insulin resistance, insulin levels are expected to increase and together elevated cortisol and insulin may promote the deposition of body fat, the formation of atherosclerotic plaques and the development of metabolic syndrome (Bjorntorp and Rosmond, 2000; Brunner *et al.*, 2002; McEwen and Wingfield, 2003). In support of this hypothesis, there is evidence that cortisol levels are raised in obese individuals (Bjorntorp and Rosmond, 2000) and that men with metabolic syndrome show higher 24-hour levels of cortisol metabolites (Brunner *et al.*, 2002).

Given the short-term effects of cortisol outlined above it is clearly also possible that HPA axis dysfunction associated with low cortisol production may lead to increased disease risk, especially if it leads to the non-suppression of other stress responses. For example, a failure to inhibit immune functions may result in increased vulnerability to auto-immune disorders, allergies and inflammation (Heim *et al.*, 2000). Hypocortisolism has been identified in patients with various illnesses, including fibromyalgia, chronic back pain and rheumatoid arthritis (Heim *et al.*, 2000), but the causality of these associations is not clear.

Finally, recent research has suggested that a disruption of the normal circadian rhythm in cortisol secretion (see below) may be associated with adverse health consequences. In study of women with breast cancer Sephton *et al.* (2000) found that women who had relatively flat diurnal cycles died significantly earlier than women with more typical cycles, while the overall level of cortisol did not predict survival.

Flattened diurnal slopes were associated with lower natural killer cell numbers and activity and the authors suggest that dysregulation of the cortisol response compromises tumor resistance by affecting immune activity.

In summary, cortisol has many complex and profound effects within the body, and disruption of the HPA axis clearly has the potential to lead to disease. It is important, then, to understand how cortisol levels vary in everyday life.

Measuring cortisol

Free versus bound hormone

A large proportion of cortisol is bound to transport proteins, such as cortisol-binding globulin and albumin, that prevent the hormone from acting on target cells. Only 2–15% of secreted cortisol circulates unbound (Kirschbaum and Hellhammer, 2000). This free hormone is responsible for the biological activity of cortisol and it is usually free cortisol that is measured in field studies.

Body fluids and different methods for collecting saliva and urine

Free cortisol can be measured in serum, urine or in saliva, using more recently developed assays. Blood collection is not usually practical for field studies since it is invasive, requires medically trained personnel, is ethically problematic in the demands it makes of participants, and can itself cause stress and thus may affect cortisol levels.

Saliva

Collection of saliva is now the preferred method, partly because the collection of saliva is less invasive than obtaining blood, and more convenient than collection of timed urine samples (Ellison, 1988; Kirschbaum and Hellhammer, 1994). These properties make it ideal for studies of infants or young children. Saliva is also safer to handle than blood or urine. Although levels of cortisol in saliva are lower than those in blood, correlations between salivary and free blood cortisol levels usually exceed 0.90 (Kirschbaum and Hellhammer, 2000).

It is now thought that saliva flow rate does not affect salivary cortisol levels, so that researchers do not need to be concerned about changes in the flow rate or composition of saliva caused by sympathetic nervous system arousal during stress (Baum and Grunberg, 1995).

Saliva can either be collected by direct salivation into small containers, or, more commonly, by the use of devices such as the Salivette. Prior to saliva collection, participants should be first asked to rinse their mouths with clean water and then examined for oral bleeding as food and blood can contaminate samples (Ellison, 1988; Flinn, 1999). Gum can be used to stimulate saliva production. Ellison (1988) has recommended the use of Wrigley's spearmint gum as the least likely to affect cortisol results. However, he cautions that comparative tests be performed by each investigator as potential problems may arise depending on the cross-reactivities of the antiserum used for sample preparation, and it is also possible that gum constituents will change over time (Ellison, 1988).

The preferred method for storage of saliva for subsequent cortisol assay is freezing, but samples can be stored for up to four weeks at 20°C without significant reduction in cortisol levels (Kirschbaum and Hellhammer, 1994). Normal mail systems can be used to send samples to laboratories. Clements and Parker (1998) simulated mailing by exposing salivary samples to fluctuating temperatures (60°−100° F) within a closed automobile. Their results suggest not only that mailing does not adversely affect results but that other situations of fluctuating temperatures, such as encountered in many field conditions, may have little impact on saliva samples. Saliva stored in cotton rolls may be subject to mold after a few days but this does not appear to impact cortisol assays (Kirschbaum, personal communication, 2005). If storage without freezing is required for longer periods a preservative such as sodium azide can be added to samples (Kirschbaum and Hellhammer, 1994; Flinn, 1999).

Equipment needed for collection of saliva

For direct collection by expectoration, a variety of inexpensive collection containers available from medical supply companies can be used. The saliva must then be pipetted into a polypropylene tube for assay. To avoid the pipetting step, some investigators have had participants spit into a test tube through a short straw (Shirtcliff *et al.*, 2001). Saliva can be collected with a dental cotton roll or a braided cotton rope either by the participant or by the investigator in the case of young children or adults with cognitive impairment. Saliva can be extracted from the cotton roll using a needleless syringe. Saliva is more commonly collected with the Salivette system (Sarstedt) which contains a dental cotton (or polyester) roll in a small tube with a hole within a standard centrifuge tube. Participants are instructed to chew on the cotton roll, which is believed to stimulate saliva flow, and then place the cotton roll in the tube which is easily centrifuged to release the saliva into the larger tube. Citric acid treated salivettes (or citric acid stimulation prior to collection) should not be used as the reduction in saliva pH can lead to false high values of cortisol with modern immuno-assay techniques (Kirschbaum and Hellhammer, 2000). Recently, several studies have suggested that the cotton rolls may interfere with the assay of some hormones (Lenander-Lumikari *et al.*, 1995; Kruger *et al.*, 1996; Shirtcliff *et al.*, 2001); however, plain cotton salivettes do not appear to impact cortisol assays (Shirtcliff *et al.*, 2001). Therefore, investigators who intend to use saliva for additional assays should not use cotton-based techniques. The direct technique is less expensive than the Salivette system; however, the Salivette system is more hygienic and probably easier to give participants who are collecting saliva on their own. Braided ropes are recommended for participants who are at risk of choking such as young children or adults with cognitive impairment (Hodgson *et al.*, 2004). Regardless of technique, investigators should also carry bottled water and small cups in the field so that participants can rinse their mouths prior to collection. Latex or other gloves are also recommended for handling samples.

Cortisol has a strong circadian pattern so time of collection must be recorded. If participants are collecting samples on their own, researchers should invest in good watches with alarms. Timex makes a watch (Timex Triathlon with Data Link®) which can be programmed with up to 100 alarms. Along with the alarm, messages to cue specific tasks (saliva collection or diary recording) can be added. Although this has not been tested, the watch will likely improve compliance. Personal digital assistants (PDAs) can also be used and may be combined with the collection of ecological momentary data (see p. 137 and Chapter 3). When working with populations which are less concerned with time or have low literacy rates, the investigator should collect the samples rather than relying on participants to follow a complex collection protocol. While the intrusion of the investigator in the daily routine of participants may lead to changes in cortisol secretion, there will be no question about compliance.

Compliance is a major issue in cortisol research. Two studies have been conducted which compare reported compliance with actual compliance as measured by an electronic monitor cap (Kudielka *et al.*, 2003;

Broderick *et al.*, 2004). Both studies reported that compliance was poor overall but was better with those who knew that compliance was being monitored than those who did not. They further noted that conclusions based on statistical analysis differed by level of compliance. Aardex Ltd. manufactures an electronic monitor cap, MEMS® Smart Cap (formally eDEM™). The price varies depending on quantity but caps cost about $100 each and associated hardware and software is approximately $1000 at the time of printing.

Urine

Urine may be collected directly into large (500 ml or 1 l) screw-top containers. The whole urination must be collected because it is necessary to know the volume of urine produced. Urine should be frozen as soon as possible after collection.

There is some inconsistency in the methods used to express urinary levels of cortisol. It is necessary to know both the time of the urination previous to collection, and the timing of collection of the urine used for assay, in order to calculate the amount of hormone excreted over a particular time period. This can be a particular problem in cultures where time is not closely monitored. Some researchers also express the rate of excretion per kilogram body weight (e.g. Frankenhaeuser *et al.*, 1989). Others have not taken body weight into account, since there appears to be no well-defined relationship between cortisol excretion rate and body weight (Lundberg and Forsman, 1980). There is, however, a correlation between urine flow rate and cortisol excretion rate (Pollard, 1995). It seems most likely that this correlation arises because of passive diffusion of hormones into the urine, suggesting that it is probably appropriate to adjust cortisol excretion rates for urine flow rate. One method of achieving this adjustment is to use the residuals resulting from the regression of cortisol excretion rate on urine flow rate.

Assay techniques

Salivary and urinary cortisol is usually assessed by radioimmunoassay (RIA). Increasingly, commercial assay kits designed for use with serum or plasma now also provide instructions for use with saliva (Kirschbaum and Hellhammer, 2000). This technique relies on the binding of antisera to cortisol. Radioisotopes are used to label the cortisol in single-antibody RIA, or to compete for binding with cortisol before measurement in competitive binding RIA. This is a simple and easy to use technique which is very reliable, sensitive and specific, and has been well validated (Baum and Grunberg, 1995). Kits for measuring serum levels of free cortisol have been available for some time and have been adapted to measure the lower levels found in saliva. Enzyme Linked ImmunoSorbent Assay (ELISA) methods are also increasingly available, avoiding the necessity of a laboratory equipped to handle radioactivity. It is also possible to use high performance liquid chromatography (HPLC), which can detect levels of several steroid hormones simultaneously, but this method is generally more expensive. Saliva specific enzyme immunoassays (EIA) kits can be purchased from Salimetrics (www.salimetrics.com) and DSL (www.DSLabs.com). Typically the cost of a salivary cortisol assay is approximately $3—4 for a duplicate assay.

In addition to the various assay kits, investigators can send samples to local medical laboratories or to national research laboratories such as Salimetrics, DSL and the laboratory of Dr. Clemens Kirschbaum. Costs vary by lab and assay technique but range from $5—15 per sample. Many labs will provide a discount for bulk orders. Investigators should consider testing the reliability of the lab. This can be accomplished by dividing up a saliva sample into multiple aliquots and examining the intra-assay coefficient of variation.

Circadian rhythm and other extraneous influences on cortisol levels

In studies assessing stress and cortisol researchers are usually interested in the direct relationship between the two. In order to pinpoint this relationship it is important to take account of other factors which may affect cortisol levels.

Circadian rhythm

Cortisol follows a strong circadian rhythm, which must be considered when making decisions about sample collection. It is secreted in intermittent pulses at 1—2 hour intervals; the circadian pattern is produced by the height of successive pulses. This pattern is established as early as three months of age (Price *et al.*, 1983). In people following normal

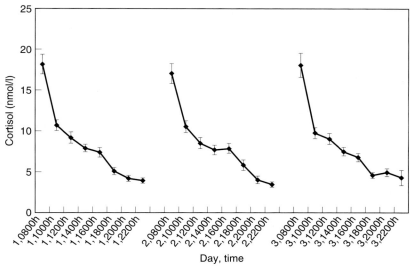

Figure 5.2.   Average cortisol levels over three days for 48 volunteers aged over 65 and living in the Twin Cities Greater Metropolitan area, Minnesota, USA. From (Ice *et al.*, 2004).

daily schedules, levels are highest in the early morning and lowest between 2000 h and 0200 h (Figure 5.2), with a small peak associated with a lunch-time meal (Weitzman *et al.*, 1971; Pruessner *et al.*, 1997; Kirschbaum and Hellhammer, 2000). As with other circadian rhythms, cortisol appears to be influenced by sleep (Born and Fehm, 1998) and light conditions (A. Levine *et al.*, 1994) and is ultimately controlled by the suprachiasmatic nucleus of the hypothalamus. ACTH and cortisol levels start to rise during late sleep and it has been hypothesized that they may help stimulate waking (Born *et al.*, 1999). Waking stimulates a further rise in cortisol, which reaches a peak around 30 minutes after waking.

Although this diurnal pattern has been considered robust, alterations have been observed in disease states such as rheumatoid arthritis (Dekkers *et al.*, 2000a,b), colic (White *et al.*, 2000) and among institutionalized children (Carlson and Earls, 1997), populations living in circumpolar environments (M. E. Levine *et al.*, 1994) and shift workers (Hennig *et al.*, 1998; Goh, 2000). Furthermore, recent studies have identified flattened cycles in healthy populations (Smyth *et al.*, 1997; Stone *et al.*, 2001) and a flattened slope is associated with an increased risk of mortality among breast cancer patients (Sephton *et al.*, 2000).

Thus, it has been suggested that diurnal cycle variation may provide valuable information on physiological and environmental influences on the HPA axis and ultimately health outcomes (Stone *et al.*, 2001).

In order to make comparisons across situations, either in the same person or in different people, samples must be carefully timed to take account of the circadian rhythm. One strategy is to collect samples at fixed times so that variation due to circadian rhythm is eliminated as much as possible. Alternatively, because waking plays an important role in the diurnal rhythm of cortisol secretion, it may be preferable to synchronize sample collection to waking (Edwards *et al.*, 2001b). Another alternative is to collect information on waking time and to correct for time since waking in analyses (Flinn, 1999). Unless the researcher is interested in measuring the waking response or circadian pattern (see below) it is usually sensible to avoid sample collection during the early morning hours, when the drop in cortisol levels is at its steepest. During this period of the day time of collection is likely to overwhelm other sources of variation in cortisol levels, as indicated by meta-analytic findings that cortisol effect sizes in response to laboratory stressors were much smaller in the morning than in the afternoon (Dickerson and Kemeny, 2004).

Season

While few studies have examined seasonal differences there is some evidence that cortisol varies by season in temperate climates (Maes *et al.*, 1997; King *et al.*, 2000; Hansen *et al.*, 2001). The pattern of variation may be different between genders (Maes *et al.*, 1997). Most studies have reported higher cortisol levels during winter months (Maes *et al.*, 1997; Walker *et al.*, 1997; King *et al.*, 2000; Hansen *et al.*, 2001). Rosmalen *et al.* (2005) reported highest levels in "mid-year" (June–August) among 10–12 year old children; however, this is based on a sampling distribution rather than repeated measures in each subject. As these studies have been conducted in Northern Europe, it has been suggested that the time of dawn or amount of daylight might influence cortisol levels (Walker *et al.*, 1997; Hansen *et al.*, 2001); however, it is equally likely that behavioral factors might influence such variation. Both activity patterns and affect are likely to change with season.

Age

Although the typical circadian pattern is believed to be established by three months of age (Price *et al.*, 1983), patterns, levels and reactivity appear to undergo several changes in early childhood (Gunnar and Donzella, 2002). Newborns have low levels of plasma (total) cortisol but salivary (free) cortisol levels are similar to those in adults (Gunnar, 1992). This is presumably due to the amount of cortisol binding globulin. Prior to the establishment of a typical morning peak, infants have two peaks, unrelated to time of day (Sippell *et al.*, 1978). At around three months, infants appear to have a dampened response of cortisol to stressors even in the presence of obvious behavioral distress (Ramsay and Lewis, 1994; Gunnar *et al.*, 1996; Larson *et al.*, 1998). This appears to last up to 18 months (Ramsay and Lewis, 1994; Gunnar *et al.*, 1996; Gunnar and Donzella, 2002). Among newborns, cortisol concentrations are low following periods of sleep compared to awake non-distressing activity (Gunnar *et al.*, 1985). Generally, among infants behavior correlates with cortisol only under stressed conditions (Larson *et al.*, 1991). Although the typical high morning and low evening levels are established early, Gunnar and collegues have found that a consistent decrease in cortisol from mid-morning to mid-afternoon only occurs after the development of mature sleep/wake cycles and is influenced by naps, car rides and daycare settings (Larson *et al.*, 1991; Tout *et al.*, 1998; Dettling *et al.*, 1999, 2000; Watamura *et al.*, 2002, 2003). Among toddlers, cortisol levels increase across the day and the reactivity in childhood appears to be affected by social skills and comfort (Gunnar *et al.*, 1997; Dettling *et al.*, 1999; Watamura *et al.*, 2003), playtime with peers (Watamura *et al.*, 2003), age of the child (Dettling *et al.*, 1999), ability to control negative affect and aggressive behavior (Tout *et al.*, 1998; Dettling *et al.*, 1999; 2000), temperament (Gunnar *et al.*, 1997) and daycare quality (Tout *et al.*, 1998; Dettling *et al.*, 2000). Gunnar and Donzella (2002) note that a rise in cortisol during daycare is most commonly found in children between ages 3–4, a time when social relations with peers become increasingly important in daycare settings. The majority of studies on children have been conducted in the USA and often in an experimental or daycare context. Flinn has conducted a naturalistic longitudinal study of children in Dominica which has added to the richness of the data on HPA responsiveness in children. He has reported that household composition, household events, gender and temperament all influence cortisol levels and reactivity among children (Flinn *et al.*, 1992; Flinn and England, 1995; Flinn *et al.*, 1995; Flinn *et al.*, 1996; Flinn and England, 1997; Flinn, 1999).

The HPA axis may become altered in old age. Evidence from animal studies indicates that with age, rodents and primates become hyper-cortisolemic, dexamethasone resistant and have a low threshold for feedback resistance (Sapolsky, 1990). The human research is more limited and equivocal. With age there may be a decreased ability of the HPA to return to baseline following a stressor, possibly due to a decreased sensitivity of glucocorticoid receptors (McEwen, 1988; Sapolsky, 1992). Nicolson *et al.* (1997) found a moderate increase in basal salivary corti-sol levels with age but a diminished reactivity with age. In a longitudinal study, Lupien *et al.* (1996) found no age associated change in basal plasma cortisol levels. Raff *et al.* (1999) found that older adults had elevated evening cortisol compared to a younger sample but there was no difference in morning levels. Others have reported a shift and an increase in the plasma cortisol nadir and/or an increase in the acrophase in plasma cortisol with age (Sherman *et al.*, 1985; Copinschi and Van Cauter, 1995; Deuschle *et al.*, 1997; Van Cauter *et al.*, 1998). In a study of community-dwelling older adults, Ice *et al.* (2004) found that healthy older adults had slightly higher levels of cortisol throughout the day compared to those reported for younger adults but that they retained normal circadian patterns (Figure 5.2).

It has been suggested that early childhood development and the HPA axis interact and that this bidirectional feedback can lead to alterations in the HPA axis and cortisol in adulthood. These suggestions have largely come from the literature on post-traumatic stress disorder (PTSD). Specifically, it has been suggested that childhood trauma or maltreatment can lead to overproduction of cortisol, which has been hypothesized to lead to HPA axis dysregulation and ultimately hypo-cortisolism. Hypocortisolism has been observed in several studies of PTSD and some chronic pain syndromes (Poteliakhoff, 1981; Griep *et al.*, 1993). The literature on the HPA response to trauma and maltreatment in childhood is equivocal, with some studies showing no difference between exposed and unexposed children while others report increases or decreases (see Cicchetti and Rogosch [2001] and Gunnar and Vazquez [2001] for good reviews). The results seem to depend on the nature of the maltreatment, the timing of the cortisol assessment relative to maltreatment and co-morbid mental health conditions. The adult PTSD and chronic pain syndrome literature is also inconsistent (Marshall and Garakani, 2002). Some of the differences in findings in both sets of studies can be attributed to methodological disparities. In the adult literature, there appears to be far less attention to behavioral variables, co-morbid conditions and sociocultural context than explored among children. However, the main problem with the theories connecting early to later HPA alteration is the lack of longitudinal research. Without longitudinal studies, it remains unclear how trauma affects HPA longterm.

Sex

Higher cortisol levels and a greater cortisol response to laboratory stressors in men have been reported in some studies (Kirschbaum *et al.*, 1992), but in Dickerson and Kemeny's (2004) meta-analysis of laboratory studies, there were no sex differences in cortisol effect sizes in response to experimental stressors. Pruessner *et al.* (1997) reported that pre-menopausal women secreted higher levels of cortisol than men in the first hour after waking, but found no difference in the waking response between post-menopausal women and men. However, in their naturalistic study of salivary cortisol in everyday life Smyth *et al.* (1998) found that men showed a larger cortisol response to stress. They were able to show that the difference did not seem to be due to gender differences in the reporting of mood or stressors. Thus gender differences should be investigated in all studies of stress and cortisol.

Pregnancy, oral contraceptive use and the menstrual cycle

Kirschbaum *et al.* (1992) reviewed available evidence and concluded that levels of cortisol are elevated during the final trimester of pregnancy. In addition to the increase in cortisol over the course of pregnancy, the cortisol response to stressors may also differ across trimesters (Obel *et al.*, 2005). The use of oral contraceptives leads to an increase in cortisol-binding globulin levels, so that more of the total cortisol is bound and less is available as free hormone (Kirschbaum and Hellhammer, 2000). Information on pregnancy status and oral contraceptive use should therefore be collected from women. Evidence of systematic variation in cortisol levels during the menstrual cycle has not been reported (Pollard, 1995).

Eating and drinking

Eating and caffeine and alcohol consumption are all potential confounders in studies of stress and cortisol. Eating and caffeine consumption cause rises in cortisol levels, peaking around 40–45 minutes after consumption (Smyth *et al.*, 1998). Plasma cortisol levels rose after alcohol administration in a number of studies, although some authors have failed to find such an effect (Prinz *et al.*, 1980). Ice *et al.* (2004) found that both food and caffeine intake prior to sample collection were associated with inconsistent diurnal cycles of cortisol secretion in a group of older adults. It is important to recognize that the effects of stress on cortisol may be mediated by stress-related changes in eating, or drinking caffeine or alcohol, and to perform appropriate statistical analyses depending on whether this effect is of interest. That is, if the researcher is interested in all mechanisms by which stress may affect cortisol level, it would be inappropriate simply to control for the effects of eating and drinking. Instead, analyses should be conducted to test whether eating and drinking might mediate or moderate some of the effects of stress on cortisol level (Baron and Kenny, 1986).

Exercise

Cortisol levels appear to rise during exercise and peak about 20–30 minutes afterwards (Kindermann *et al.*, 1982; Kirschbaum and Hellhammer, 1994; Filaire *et al.*, 1996; Jacks *et al.*, 2002; Tremblay *et al.*, 2004). Kirschbaum *et al.* (1992) suggest that physical exercise only has an effect on salivary cortisol at levels greater than 70% VO_2 max. Thus it is important to collect information on the intensity and timing of exercise. Studies examining change in cortisol secretion after extended training have equivocal results (Lucia *et al.*, 2001; Buyukyazi *et al.*, 2003); however, most research on exercise and cortisol includes relatively intense exercise protocols primarily among trained athletes. We know little about the impact of daily physical activity on cortisol levels. Ice (in press) found that higher activity levels, measured by the Baecke Questionnaire of Habitual Physical Activity for Older Adults, were associated with a steeper decline in cortisol levels over the day.

Smoking

Nicotine causes increased levels of cortisol (Kirschbaum and Hellhammer, 1989). Wüst *et al.* (1992) found that smoking two cigarettes resulted in elevation of salivary cortisol levels, peaking at 25–35 minutes after smoking. Smoking may be related to stress, so, as with eating and drinking, statistical control for smoking must be undertaken with care.

Medication

Patients being treated with glucocorticoids should normally be excluded from studies of cortisol variation.

Sampling frequency and timing

Salivary and urinary free cortisol levels reflect cortisol secretion over different time periods and it is important to use the appropriate medium for any particular study (Baum and Grunberg, 1995; Pollard, 1995).

Saliva samples

Peak concentrations in saliva are observed 1–2 minutes after maximal concentrations in blood. It is therefore possible to collect detailed information on individual variation in salivary cortisol levels over the course of a few hours or days. Field studies can make use of the experience sampling method or ecological momentary assessment (Shiffman and Stone, 1998), where repeated samples are collected during normal everyday life (see Chapter 3). As noted above, participants can be prompted to collect data by beeps from a pre-programmed wristwatch.

Sampling protocols vary widely across studies and there is no consensus on the timing of collection, number of samples per day or the number of sampling days required for a good study. In 1999, at a meeting organized by the MacArthur Research Network on SES and Health, several leading cortisol and stress researchers discussed methodological issues associated with salivary cortisol measurement (Stewart and Seeman, 2000). The participants noted (and we concur) that the research question must drive the chosen sampling protocol. The group, however, came to the consensus that the more measurements per day, the better.

As noted previously, compliance with collection can be poor (Kudielka *et al.*, 2003; Broderick *et al.*, 2004). Investigators should investigate the use of electronic monitors to measure compliance with sample collection. While this may not always be practical, it is clear that researchers need to do everything they can to ensure compliance with sample collection regimes.

A variety of sample collection protocols has been used to investigate differences in average cortisol levels between individuals. The average of several samples collected over the course of a day has been used to provide an indication of between-individual variation in overall cortisol levels (e.g. Edwards *et al.*, 2001a). Alternatively the area under the curve for one or several days of assessment can be calculated. At the MacArthur Research Network meeting (see above) Cohen and Schwartz suggested that the minimum number of samples per day is 4–5 with ideal time of measurement being 1, 4, 8 and 11 hours after wakening to get a good estimate of area under the curve. One-day assessment was described as a "weak approach." Schwartz suggested that 3–4 days are required to have an appropriate assessment of a "trait" daily level. However, these are clearly ideal protocols, requiring sufficient funding and appropriate field conditions. Furthermore, the MacArthur Network went on to settle on a six-sample, one-day protocol for its large epidemiological study.

Given that differences in cortisol levels between individuals will be influenced by many factors other than stress, and the power that repeated saliva samples offer for examining within-individual variation in cortisol levels, more interest has been focused on investigating how diurnal variation and other within-individual variation in cortisol levels may be affected by stress. It has been suggested that absence of the typical diurnal pattern of cortisol release could indicate dysregulation of the HPA axis and some researchers have therefore been interested in the extent of the cortisol decline over the day. This has been examined by calculating a difference score between an initial waking sample and one taken at the end of the day (Edwards *et al.*, 2001a), by regressing cortisol level against time to calculate the slope of the change in cortisol over the day (Sephton *et al.*, 2000; Ice *et al.*, 2004) or by modeling individual variation in diurnal cortisol slopes using multilevel modeling (Stone *et al.*, 2001). Using this last technique Stone *et al.* have demonstrated significant between-individual variation in diurnal cortisol slopes. Unfortunately methods to assess diurnal variation have not been standardized and the method chosen may have an important impact on findings (Ice *et al.*, 2004).

Some researchers have become interested in what is known as the waking response, defined as the change in cortisol over the first 30−60 minutes of the day. As noted above, morning wakening is followed by brief ACTH and cortisol pulses (Pruessner *et al.*, 1997), with the increase in salivary cortisol peaking about 30 minutes after waking. One method used to calculate the waking cortisol response is to ask participants to collect saliva immediately upon waking and then 15, 30 and 45 minutes later, although sometimes samples have been collected for an hour after waking and at slightly different intervals. If using the first strategy, the mean increase of the last three samples over the level on the first waking sample can be used as a measure of the waking cortisol response (Wüst *et al.*, 2000; Edwards *et al.*, 2001a). Alternatively it is possible to calculate the area under the cortisol curve (AURC) with reference to the first waking sample to provide a measure of the extent of the cortisol response to waking. Another measure which can be calculated from the same information is the area under the curve (AUC) for all four waking samples, which simply provides a measure of the overall level of cortisol during the hour after waking.

$$AURC = sample2 + sample3 - (2 * sample1) + ([sample4 - sample2]/2)$$

$$AUC = sample1 + sample2 + sample3 + ([sample4 - sample1]/2)$$

Pruessner *et al.* (1997) found that the level of cortisol in the hour after waking, as measured by the area under the curve for repeated samples during that hour (AUC), had moderate to high within-person stability and suggested that it provides a reliable assessment of between-person differences in cortisol secretion. Schmidt-Reinwald *et al.* (1999) found a positive correlation between this measure and the cortisol response to infused ACTH, and suggest AUC therefore reflects the capacity of the individual's adrenal cortex to respond to stimulation. In addition, Edwards *et al.* (2001b) found that this same measure of cortisol levels in the hour after waking predicted mean cortisol levels throughout the following 12 hours. The cortisol response to waking (as measured by the AURC) does not appear to be related to overall levels during the rest of the day (Edwards *et al.*, 2001b).

A final study design uses repeated salivary cortisol assessment in conjunction with repeated diary reports of activities and feelings to investigate the influence of daily experience on cortisol levels. Concurrent collection of data in this way avoids problems associated with the retrospective reporting of experiences and feelings (Schwartz and Stone, 1998). Given the time it takes for cortisol levels to respond it is preferable to ask participants to collect saliva 20 minutes after completing their diary (Smyth *et al.*, 1998). Diaries can be completed electronically using a hand-held computer or PDA, which has the added advantage of producing directly downloadable data and also of verifying the time at which participants entered their data (Kamarck *et al.*, 1998). Diaries will not be feasible in many studies due to illiteracy or cognitive limitations. In these circumstances behavioral observation can be used to substitute diaries. Diaries and behavioral observation are discussed in detail in Chapter 3. Again, multilevel modeling offers the most powerful method of analysis of such data, since it allows within-individual effects to be statistically separated from between-individual effects (Schwartz and Stone, 1998). It also allows the use of all data points rather than requiring the exclusion of participants with any missing data as in traditional analytical techniques. Analysis of repeated measures is further discussed in Chapter 9.

Urine samples

Hormone assayed in urine usually represents a pooling of levels over a period of hours since hormone excreted in urine reflects hormone secreted some time before, and is also an integration of the hormone excreted into the urine since the previous urination. Thus less "noise" is caused by minute-to-minute variation in serum cortisol levels when using urine samples. It is therefore appropriate to collect urine when cortisol secretion over longer time periods is of interest. Timed urine samples, e.g. over a two-hour period as in Pollard *et al.* (1996), provide a useful method for comparing cortisol levels between individuals using only one sample per individual, or, more powerfully (see above) for looking at within-individual variation over longer time periods, such as days. (Multiple saliva samples can provide similar information.) Evidence for the time lag between cortisol secretion and its excretion in urine is sparse, but it is likely to be around three hours (Fibiger *et al.*, 1984; Jenner, 1985) and allowance must therefore be made for this time lag.

It is possible to investigate levels of urinary cortisol for even longer periods of time by asking people to make 24-hour or overnight urine collections, as has been done in anthropological investigations of catecholamine variation (Brown, 1982; Jenner, 1985). However, it is not always possible to be sure that samples are complete for the time of interest, and anecdotes indicate that researchers have sometimes been suspicious about the "urine" samples provided by study participants.

Cortisol as a "stress hormone"

Cortisol has been used as a marker of stress because it is often elevated in response to psychosocial stress. Several studies have demonstrated that cortisol increases with laboratory "stressors." However, these studies provide limited information about the effect of daily stressors on cortisol and ultimately health. The data from naturalistic studies on daily stressors and cortisol are less consistent and seem to vary by a number of individual and population level characteristics (Pollard, 1995).

Laboratory studies

For many years research on cortisol variation in humans was mainly conducted in laboratory settings, or sometimes in the context of real-life situations which approximated an experimental design, such as examinations. The focus of these studies has been the measurement of hormone response to situations and stimuli assumed to be stressful. Mason (1968) reviewed the earliest work of this type, in which levels of a biochemical derivative of cortisol, 17-hydroxycorticosteroid (17-OHCS), were measured. Levels were shown to rise in people performing laboratory tasks, such as mental arithmetic, watching disturbing films and taking examinations. Mason concluded that psychological influences have powerful effects on HPA activity and that elevation of 17-OHCS appeared to reflect a state of emotional arousal. It was suggested that novelty, uncertainty, and unpredictability were particularly potent in eliciting 17-OHCS elevation. Later work showed that levels of cortisol itself increased over baseline in men performing cognitive-conflict tasks (Collins and Frankenhaeuser, 1978), doing mental arithmetic (Holl *et al.*, 1984), or performing memory and multiple-choice tasks (Wittersheim *et al.*, 1985). Relatively few of these early studies included women.

The results of some of these studies formed the basis of the influential model of stress hormone variation formulated by Frankenhaeuser and colleagues (Lundberg and Frankenhaeuser, 1980). They suggested that negative affect (distress) is the main psychosocial determinant of increased cortisol secretion. The effects of distress were distinguished from the effects of effort or positive psychosocial arousal, which were considered to influence catecholamine but not cortisol levels. Henry (1982) made similar proposals, also linking cortisol secretion with distress and lack of control. In probably the best experimental test of the different psychophysiological effects of effort versus distress, Peters *et al.* (1998)

found that cortisol levels in men did not rise in tasks designed to induce effort and accompanied by a controllable noise, but did rise during the same tasks when the noise was made uncontrollable. Both conditions were subjectively rated as aversive, but those which were uncontrollable were more aversive. These results appear to support the Frankenhaeuser model.

Recently, Dickerson and Kemeny (2004) performed a meta-analysis of laboratory studies of the effects of acute psychological stressors on cortisol levels. They found that the stressors which had the biggest effects were those that involved a social-evaluative threat, when others could negatively judge performance, and which were also uncontrollable. Social evaluation and uncontrollability had independent effects on cortisol response. Social-evaluative, uncontrollable tasks were also associated with a more prolonged cortisol elevation. They were not, however, associated with greater increases in subjective distress than other types of stressors and they found no relationship between reported distress and cortisol. Dickerson and Kemeny suggest, therefore, that the experience of distress is not sufficient to elicit a cortisol response. They propose that stressors which threaten social status, particularly when that threat is augmented by uncontrollable conditions, cause feelings of shame and related emotions and that these emotions could lead to the production of cortisol. Dickerson and Kemeny note that their findings highlight the importance of considering the social context when examining cortisol responses.

Laboratory studies have traditionally focused on aversive stimuli and most have not assessed emotional responses to the stimuli. However, there are now several studies which have explicitly tested the role of emotion as the primary mediator of the cortisol response, by exposing participants to both negative and positive stimuli and measuring their subjective experience as well as their cortisol response. Codispoti *et al.* (2003) report cortisol elevations in men after viewing pictures which evoked negative affect but not after viewing pictures which evoked positive affect in the form of sexual arousal. Buchanan *et al.* (1999) found a decline in cortisol levels when viewing video clips which reduced negative affect at the same time as increasing positive affect. In contrast, Brown *et al.* (1993) found that cortisol levels rose when they induced feelings of either sadness or elation in a group of women. Notably, the types of positive affect induced in these studies were very different and it may be more difficult to induce strong feelings of positive affect than of negative affect in an experimental situation. In general, experimental studies strongly suggest that negative affect is associated with a rise in cortisol levels, while the role of positive affect is less well understood. Questions also remain about the applicability of findings from studies of artificially induced stress, or extreme experiences in real life, to cortisol variation in everyday life.

Naturalistic studies ⁁

Fewer studies have been conducted in naturalistic settings although such studies are becoming more common with the increased availability of salivary cortisol assays. Many of the early studies of cortisol in naturalistic settings were conducted among people in "high stress" jobs (e.g. air traffic controllers, emergency medical technicians) (Hytten *et al.*, 1990; Zeier, 1994; Zeier *et al.*, 1996) or people participating in high risk activities (e.g. bungee jumping) (Hennig, 1994), and found the expected cortisol elevations. Cortisol has also been found to increase in a number of controlled real-life situations including during exams (Frankenhaeuser *et al.*, 1978), dental examinations (Benjamins *et al.*, 1992; Miller *et al.*, 1995), smoke diving (Hennig, 1994), and on rowing competition days (Pearson *et al.*, 1995).

Many naturalistic studies have focused on the effects of paid work on cortisol levels but the findings are not easy to interpret (see Pollard, 1995 for review). In general, studies have not shown that cortisol levels are elevated at work compared to leisure (e.g. Pollard *et al.*, 1996). One study to find an effect of work showed that in air traffic controllers there was a positive correlation between mental workload and change in cortisol levels from baseline (Zeier *et al.*, 1996). However, many of these studies did not collect data on affect and without knowing how work influences mood it is difficult to interpret the data. For example, in the case of mothers, it has been suggested that women who work outside the home may experience less stress during the day than women who stay at home to look after children. Furthermore, many of these studies used research designs with sub-optimal sampling of cortisol levels.

Several studies have examined the impact of self-reported daily hassles and emotion measured using diaries or ecological momentary data (EMA). Generally, these studies find that cortisol increases with negative affect and daily hassles which are assessed as stressful (e.g. Smyth *et al.*, 1998). EMA studies of cortisol include multiple sampling points. Sometimes the sampling is at a fixed point and combined with diary recording of mood and/or stressful events; others have signaled collection of cortisol and diary recordings separately at random points.

Clearly, these methodological differences can result in different findings across studies, but in general these study designs appear the most likely to identify links between daily experiences and cortisol levels.

Using such a design Ockenfels *et al.* (1995) found no difference in cortisol levels between unemployed and employed study participants, but perceived stress as measured by the Perceived Stress Scale (PSS) predicted cortisol levels in both groups. Using a similar methodology with all working adults, van Eck *et al.* (1996) found that cortisol increased with trait anxiety, agitation (state), negative affect (state) and stressful events. Stressful events had a greater effect on cortisol if they were ongoing at the time of cortisol collection, and frequently experienced stressful events had a smaller effect. Smyth *et al.* (1998) similarly found that cortisol increased with current stressful events and anticipated stressful events. Cortisol was higher when events were rated as more stressful. Neither controllability nor previous experience with the event predicted cortisol. Cortisol was found to decrease with positive affect and increase with negative affect. Hanson *et al.* (2000) reported that among a group of employed health professionals and clerks, negative mood state was associated with increasing cortisol, whereas negative mood trait was not. In a study of older adults, Ice (in press) found that positive affective trait was associated with lower cortisol levels while positive mood state was associated with higher cortisol levels. Negative mood state and perceived stress (trait) were also associated with cortisol in bivariate analyses. In multivariable analyses, however, perceived stress was the only affect trait that remained significant and no state variables were significant. Overall, the effect sizes of trait variables were higher, suggesting that a larger sample size is needed to assess the impact of mood states. Using a different methodology, Decker (2000) reported that frequency of distressed mood over 15 days was positively associated with higher cortisol level. Frequency of reported stressors was not associated with cortisol. Because the mood data were collected by an observer, the author suggested that there may have been an under-reporting of social stressors.

In a comparison between depressed and non-depressed participants, Peeters *et al.* (2003) found that negative events were not associated with a change in cortisol levels among depressed individuals. Negative affect was associated with increased cortisol, although to a lesser extent, among depressed participants. In addition, there was an interaction between negative events and gender, such that depressed women showed a larger increase in cortisol in response to negative events than depressed men.

It is clear that the power of EMA studies is beginning to uncover fine-grained associations, particularly between mood and cortisol levels, which were obscured with previous methodologies. There is now evidence that cortisol levels rise with negative affect and decline with positive affect, although even here findings are complex and not always easy to interpret. One area which requires further research is the effect of habituation to generally stressful experiences such as work on cortisol responses. This question is related to the issue of chronic stress, as explored below.

Chronic stress and the HPA axis

The effect of chronic stress on the HPA axis and cortisol levels has recently received a great deal of attention. This area of research developed out of the seemingly paradoxical observation, noted above, that individuals experiencing certain kinds of chronic psychosocial stressors or chronic pain syndromes had lower cortisol levels than "unstressed" individuals (Heim *et al.*, 2000). Heim *et al.* (2000) have suggested that chronic stress may cause depletion of HPA axis hormones, hypersecretion of corticotrophin releasing factor or increased feedback sensitivity. As noted earlier, the causality of the connection between chronic stressors and hypocortisolism in these patients is unclear and may be influenced by any number of confounding variables. Moreover, decreased cortisol is not consistently observed in chronic pain syndromes. Others have reported cortisol increases in patients with fibromyalgia and rheumatoid arthritis (Catley *et al.*, 2000) and in chronic migraine sufferers (Peres *et al.*, 2001). Heim *et al.* (2000) also point to research conducted during the Vietnam war of special forces (Bourne *et al.*, 1968) and medics (Bourne *et al.*, 1967) to support their argument. Both of these studies found that most of the participants had lower than pre-dicted cortisol levels. However, all participants had effective coping mechanisms of denial, detachment or an overriding sense of altruism. Another study, which is often cited in this literature, is a study of parents of children with cancer (Friedman *et al.*, 1963). Although this study found unexpectedly low cortisol levels, the parents with the lowest cortisol levels had the strongest sense of denial. Thus the connection between some types of long-term experiences and low levels of cortisol requires greater investigation.

This research also begs the question, what is chronic stress? Very often researchers categorize people into stressed and non-stressed groups based on a priori assumptions of what should be stressful. People who we might expect to experience chronic stress may only feel stressed periodically or the stress they experience may be at a very low level.

Connecting cortisol to outcomes

As cortisol has wide-ranging effects on many systems, the number of potential outcome variables is almost limitless. Animal studies have been able to examine the impact of stress-related elevated cortisol on a number of systems including the brain, immune function and endocrine function (with a particular focus on growth, glucose metabolism and reproductive function). Due to the invasive nature of many of these types of studies and the length of time required for others, health outcomes and physiological function are less commonly examined among humans.

As the field of cortisol physiology matures, researchers are still grappling with methodology. It is still unclear what has the most significant impact on health: cortisol level, slope, pattern or consistency of pattern. To date few outcome studies have been conducted. Ultimately, the best measure of cortisol variability for health-related purposes must be determined through prospective study of morbidity and mortality outcomes.

Summary

Studies of variation in cortisol levels, and the causes and consequences of this variation, have developed significantly over the last half century. The methods used in cortisol studies must be very carefully considered, particularly because of the strong circadian pattern shown by cortisol. Nevertheless, salivary assays and ecological momentary assessment data collection techniques in particular have opened up new possibilities for powerful research designs, which need not involve great expense.

Resources

Labs

Diagnostic Systems Laboratory, Inc. (DSL)
www.DSLabs.com
800-231-7970

Dr. Clemens Kirschbaum
Department of Psychology
Dresden University of Technology
Zellescher Weg 17, A220
D-01062 Dresden, Germany
phone: +49-351-4633-9660
fax: +49-351-4633-7274
email: ck@biopsych.tu-dresden.de
internet: http://biopsychologie.tu-dresden.de

Salimetrics, LLC
P.O. Box 395
State College, PA 16804-0395
800-790-2258 (USA only) 814-234-7748
sales@salimetrics.com
www.salimetrics.com

Assay kits

Diagnostic Systems Laboratory, Inc. (DSL)
www.DSLabs.com
800-231-7970

Salimetrics, LLC
P.O. Box 395
State College, PA 16804-0395
800-790-2258 (USA only) 814-234-7748
sales@salimetrics.com
www.salimetrics.com

Equipment/supplies

Smart Caps
APREX, a division of AARDEX
2849-B Whipple Road
Union City, CA 94587
510-476-1940
http://www.aardex.ch/

Salivettes
Sarstedt
1025, St. James Church Road

P.O. Box 468
Newton NC 28658-0468
USA
Tel.: +1 828 465 4000
Fax: +1 828 465 0718
Email: info@sarstedt.com
SARSTEDT Ltd.
68 Boston Road
Beaumont Leys
Leicester LE4 1AW
United Kingdom
Tel.: +44 1162 359023
Fax: +44 1162 366099
Email: info@sarstedt.com
http://www.sarstedt.com/php/main.php?newlanguage=en
(for locations in other countries)

References

Baron, R. and Kenny, D. (1986). The moderator—mediator distinction in social psychological research: conceptual, strategic, and statistical considerations. **51**, 1173–82.

Baum, A. and Grunberg, N. (1995). Measurement of stress hormones. In *Measuring Stress: A Guide for Health and Social Scientists*, ed. S. Cohen, R. Kessler and L. Gordon. New York: Oxford University Press, 175–92.

Benjamins, C., Asschenman, H. and Schuurs, A. (1992). Increased salivary cortisol in severe dental anxiety. *Psychophysiology*, **29**, 302–5.

Berne, R. and Levy, M. (1993). *Physiology*. St Louis, Missouri: Mosby.

Bilbo, S. D., Dhabhar, F. S., Viswanathan, K. *et al.* (2002). Short day lengths augment stress-induced leukocyte trafficking and stress-induced enhancement of immune function. *Proceedings of National Academy of Sciences*, **99**(6), 4067–72.

Bjorntorp, P. and Rosmond, P. (2000). Obesity and cortisol. *Nutrition*, **16**(10), 924–36.

Born, J. and Fehm, H. L. (1998). Hypothalamus—pituitary—adrenal activity during human sleep: a coordinating role for the limbic hippocampal system. *Experimental and Clinical Endocrinology and Diabetes*, **106**(3), 153–63.

Born, J., Hansen, K., Marshall, L., Molle, M. and Fehm, H. (1999). Timing the end of nocturnal sleep. *Nature*, **397**, 29–30.

Bourne, P. G., Rose, R. M. and Mason, J. W. (1967). Urinary 17-OHCS levels. Data on seven helicopter ambulance medics in combat. *Archives of General Psychiatry*, **17**(1), 104–10.

(1968). 17-OHCS levels in combat. Special forces "A" team under threat of attack. *Archives of General Psychiatry*, **19**(2), 135–40.

Broderick, J. E., Arnold, D., Kudielka, B. M. and Kirschbaum, C. (2004). Salivary cortisol compliance: comparison of patients and healthy volunteers. *Psychoneuroendocrinology*, **29**, 636–50.

Brown, D. (1982). Physiologial stress and culture change in a group of Filipino-Americans: a preliminary investigation. *Annals of Human Biology*, **9**, 553–63.

Brown, E., Varghese, F. and McEwen, B. (2004). Association of depression with medical illness: does cortisol play a role? *Biological Psychiatry*, **55**, 1–9.

Brown, W., Sirota, A., Niaura, R. and Engebretson, T. (1993). Endocrine correlates of sadness and elation. *Psychosomatic Medicine*, **55**, 458–67.

Brunner, E., Hemingway, H., Walker, B. *et al.* (2002). Adrenocortical, autonomic, and inflammatory causes of the metabolic syndrome. *Circulation*, **106**, 2659–65.

Brunner, E. and Marmot, M. (1999). Social organization, stress, and health. In *Social Determinants of Health*, ed. M. Marmot and R. Wilkinson. Oxford: Oxford University Press, 17–43.

Buchanan, T., al'Absi, M. and Lovallo, W. (1999). Cortisol fluctuates with increases and decreases in negative affect. *Psychoneuroendocrinology*, **24**, 227–41.

Buyukyazi, G., Karamizrak, S. O. and Islegen, C. (2003). Effects of continuous and interval running training on serum growth and cortisol hormones in junior basketball players. *Acta Physiologica Hungarica*, **90**(1), 69–79.

Carlson, M. and Earls, F. (1997). Psychological and neuroendocrinological sequelae of early social deprivation in institutionalized children in Romania. *Annals of the New York Academy of Sciences*, **807**, 419–28.

Catley, D., Kaell, A. T., Kirschbaum, C. and Stone, A. A. (2000). A naturalistic evaluation of cortisol secretion in persons with fibromyalgia and rheumatoid arthritis. *Arthritis Care and Research*, **13**(1), 51–61.

Cicchetti, D. and Rogosch, F. A. (2001). Diverse patterns of neuroendocrine activity in maltreated children. *Development and Psychopathology*, **13**(3), 677–93.

Clements, A. D. and Parker, C. R. (1998). The relationship between salivary cortisol concentrations in frozen versus mailed samples. *Psychoneuroendocrinology*, **23**(6), 613–16.

Codispoti, M., Gerra, G., Montebarocci, O. *et al.* (2003). Emotional perception and neuroendocrine changes. *Psychophysiology*, **40**, 863–8.

Collins, A. and Frankenhaeuser, M. (1978). Stress responses in male and female engineering students. *Journal of Human Stress*, **4**, 43–8.

Copinschi, G. and Van Cauter, E. (1995). Effects of ageing on modulation of hormonal secretions by sleep and circadian rhythmicity. *Hormone Research*, **43**(1–3), 20–4.

Decker, S. A. (2000). Salivary cortisol and social status among Dominican men. *Hormones and Behavior*, **38**(1), 29–38.

Dekkers, J. C., Geenen, R., Godaert, G. L., van Doornen, L. J. and Bijlsma, J. W. (2000a). Diurnal rhythm of salivary cortisol levels in patients with recent-onset rheumatoid arthritis. *Arthritis and Rheumatism*, **43**(2), 465–7.

  (2000b). Diurnal courses of cortisol, pain, fatigue, negative mood, and stiffness in patients with recently diagnosed rheumatoid arthritis. *International Journal of Behavioral Medicine*, **7**(4), 353–71.

Dettling, A. C., Gunnar, M. R. and Donzella, B. (1999). Cortisol levels of young children in full-day childcare centers: relations with age and temperament. *Psychoneuroendocrinology*, **24**(5), 519–36.

Dettling, A. C., Parker, S. W., Lane, S., Sebanc, A. and Gunnar, M. R. (2000). Quality of care and temperament determine changes in cortisol concentrations over the day for young children in childcare. *Psychoneuroendocrinology*, **25**(8), 819–36.

Deuschle, M., Gotthardt, U., Schweiger, U. *et al.* (1997). With aging in humans the activity of the hypothalamus–pituitary–adrenal system increases and its diurnal amplitude flattens. *Life Sciences*, **61**(22), 2239–46.

Dickerson, S. and Kemeny, M. (2004). Acute stressors and cortisol responses: a theoretical integration and synthesis of laboratory research. *Psychological Bulletin*, **130**, 355–91.

Edwards, S., Clow, A., Evans, P. and Hucklebridge, F. (2001a). Exploration of the awakening cortisol response in relation to diurnal cortisol secretory activity. *Life Sciences*, **68**(18), 2093–103.

Edwards, S., Evans, P., Hucklebridge, F. and Clow, A. (2001b). Association between time of awakening and diurnal cortisol secretory activity. *Psychoneuroendocrinology*, **26**, 613–22.

Ellison, P. (1988). Human salivary steroids: methodological considerations and applications in physical anthropology. *Yearbook of Physical Anthropology*, **31**, 115–42.

Fibiger, W., Singer, G., Miller, A., Armstrong, S. and Datar, M. (1984). Cortisol and catecholamines changes as functions of time-of-day and self-reported mood. *Neuroscience Behavioral Reviews*, **8**, 523–30.

Filaire, E., Duche, P., Lac, G. and Robert, A. (1996). Saliva cortisol, physical exercise and training: influences of swimming and handball on cortisol concentrations in women. *European Journal of Applied Physiology and Occupational Physiology*, **74**, 274–8.

Flinn, M. V. (1999). Family environment, stress and health during childhood. In *Hormones, Health and Behavior*, ed. C. Panter-Brick and C. Worthman. Cambridge: Cambridge University Press.

Flinn, M. V. and England, B. G. (1995). Family environment and childhood stress. *Current Anthropology*, **36**, 854–66.

  (1997). Social economics of childhood glucocorticoid stress response and health. *American Journal of Physical Anthropology*, **102**, 33–54.

Flinn, M. V., England, B. G. and Beer, T. (1992). Health conditions and corticosteroid stress response among children in a rural Dominican village. *American Journal of Physical Anthropology*, **7**, 122.

Flinn, M. V., Quinlan, M., Quinlan, R., Turner, M. and England, B. G. (1995). Glucocorticoid stress response, immune function, and illness among children in a rural Caribbean village. *American Journal of Human Biology*, **7**, 122

Flinn, M. V., Turner, M., Quinlan, R., Decker, S. and England, B. G. (1996). Male–female differences in effects of parental absence on glucocorticoid stress reponses. *Human Nature*, **7**, 125–62.

Frankenhaeuser, M., Lundberg, U., Fredrikson, M. *et al.* (1989). Stress on and off the job as related to sex and occupational status in white-collar workers. *Journal of Organizational Behavior*, **10**, 321–46.

Frankenhaeuser, M., von Wright, M., Collens, A. *et al.* (1978). Sex differences in psychoneuroendocrine reactions to examination stress. *Psychosomatic Medicine*, **47**, 313–19.

Friedman, S. B., Mason, J. and Hamburg, D. (1963). Urinary 17-hydroxycorticosteroid levels in parents of children with neoplastic disease. A study of chronic psychological stress. *Psychosomatic Medicine*, **25**, 364–76.

Goh, V. H. (2000). Circadian disturbances after night-shift work onboard a naval ship. *Military Medicine*, **165**(2), 101–5.

Griep, E. N., Boersma, J. W. and de Kloet, E. R. (1993). Altered reactivity of the hypothalamic–pituitary–adrenal axis in the primary fibromyalgia syndrome. *Journal of Rheumatology*, **20**(3), 469–74.

Gunnar, M. R. (1992). Reactivity of the hypothalamic–pituitary–adrenocortical system to stressors in normal infants and children. *Pediatrics*, **90**(3 Pt 2), 491–7.

Gunnar, M. R., Brodersen, L., Krueger, K. and Rigatuso, J. (1996). Dampening of adrenocortical responses during infancy: Normative changes and individual differences. *Child Development*, **67**(3), 877–89.

Gunnar, M. R. and Donzella, B. (2002). Social regulation of the cortisol levels in early human development. *Psychoneuroendocrinology*, **27**(1–2), 199–220.

Gunnar, M. R., Malone, S., Vance, G. and Fisch, R. O. (1985). Quiet sleep and levels of plasma cortisol during recovery from circumcision in newborns. *Child Development*, **5**, 824–34.

Gunnar, M. R., Tout, K., de Haan, M., Pierce, S. and Stansbury, K. (1997). Temperament, social competence, and adrenocortical activity in preschoolers. *Developmental Psychobiology*, **31**(1), 65–85.

Gunnar, M. R. and Vazquez, D. M. (2001). Low cortisol and a flattening of expected daytime rhythm: potential indices of risk in human development. *Development and Psychopathology*, **13**(3), 515–38.

Hansen, A. M., Garde, A. H., Skovgaard, L. T. and Christensen, J. M. (2001). Seasonal and biological variation in urinary epinephrine, norepinephrine, and cortisol of healthy women. *Clinica Chimica Acta*, **309**, 25–35.

Hanson, E. K. S., Maas, C. J. M., Meijman, T. F. and Godaert, G. L. R. (2000). Cortisol secretion throughout the day, perceptions of the work environment, and negative affect. *Annals of Behavioral Medicine*, **22**(4), 316–24.

Heim, C., Ehlert, U. and Hellhammer, D. H. (2000). The potential role of hypocortisolism in the pathophysiology of stress-related bodily disorders. *Psychoneuroendocrinology*, **25**(1), 1–35.

Hennig, J. (1994). Biopsychological changes after bungee jumping: beta-endorphin immunoreactivity as a mediator of euphoria? *Neuropsychobiology*, **29**, 28–32.

Hennig, J., Kieferdorf, P., Moritz, C., Huwe, S. and Netter, P. (1998). Changes in cortisol secretion during shiftwork: implications for tolerance to shiftwork? *Ergonomics*, **41**(5), 610–21.

Henry, J. (1982). The relation of social to biological processes in disease. *Social Science and Medicine*, **16**, 369–80.

Hodgson, N., Freedman, V. A., Granger, D. A. and Erno, A. (2004). Biobehavioral correlates of relocation in the frail elderly: salivary cortisol, affect and cognitive function. *Journal of the American Geriatrics Society*, **52**, 1856–62.

Holl, R., Fehm, H. L., Voigt, K. H. and Teller, W. (1984). The "midday surge" in plasma cortisol induced by mental stress. *Hormone and Metabolic Research*, **16**(3), 158–9.

Hytten, K., Jensen, A. and Skauli, G. (1990). Stress inoculation training for smoke divers and free fall lifeboat passengers. *Aviation Space and Environmental Medicine*, **61**(11), 983–8.

Ice, G. H. (in press). Factors influencing cortisol level and slope among community dwelling older adults in Minnesota. *Journal of Cross-Cultural Gerontology*, in press.

Ice, G., Katz-Stein, A., Himes, J. H. and Kane, R. L. (2004). Diurnal cycles of salivary cortisol in older adults. *Psychoneuroendocrinology*, **29**(3), 355–70.

Jacks, D. E., Sowash, J., Anning, J., McGloughlin, T. and Andres, F. (2002). Effect of exercise at three exercise intensities on salivary cortisol. *Journal of Strength and Conditioning Research*, **16**(2), 286–9.

Jenner, D. (1985). Population studies of variation in catecholamine and corticosteroid variation. *PhD thesis*, University of Oxford.

Kamarck, T., Shiffman, S., Smithline, L. *et al.* (1998). The diary of ambulatory behavioral states: a new approach to the assessment of psychosocial influences on ambulatory cardiovascular activity. In *Technology and Methods in Behavioral Medicine*, ed. D. Krantz and A. Baum. New Jersey: Lawrence Erlbaum Associates.

Kindermann, W., Schnable, A., Schmitt, W. *et al.* (1982). Catecholamines, growth hormone, cortisol, insulin and sex hormones in anaerobic and aerobic exercise. *European Journal of Applied Physiology*, **49**, 389–99.

King, J. A., Rosal, M. C., Ma, Y. S. *et al.* (2000). Sequence and seasonal effects of salivary cortisol. *Behavioral Medicine*, **26**(2), 67–73.

Kirschbaum, C. and Hellhammer, D. (1989). Salivary research in psychobiology research: an overview. *Neuropsychobiology*, **22**, 150–69.

   (1994). Salivary cortisol in psychoneuroendocrine research: Recent developments and applications. *Psychoneuroendocrinology*, **19**, 313–33.

   (2000). Salivary cortisol. In *Encyclopedia of Stress*, ed. G. Fink. San Diego, CA: Academic Press, pp. 379–83.

Kirschbaum, C., Wuest, S. and Hellhammer, D. (1992). Consistent sex differences in cortisol responses to psychological stress. *Psychosomatic Medicine*, **54**, 648–57.

Kruger, C., Brueunig, U., Biskupek-Sigward, J. and Door, H. G. (1996). Problems with salivary 17-hydroxyprogesterone determinations using the Salivette device. *European Journal of Clinical Biochemistry*, **34**(11), 926–9.

Kudielka, B. M., Broderick, J. E. and Kirschbaum, C. (2003). Compliance with saliva sampling protocols: electronic monitoring reveals invalid cortisol daytime profiles in noncompliant subjects. *Psychosomatic Medicine*, **65**(2), 313–19.

Larson, M. C., Gunnar, M. R. and Hertsgaard, L. (1991). The effects of morning naps, car trips, and maternal separation on adrenocortical activity in human infants. *Child Development*, **62**(2), 362–72.

Larson, M. C., White, B. P., Cochran, A., Donzella, B. and Gunnar, M. (1998). Dampening of the cortisol response to handling at 3 months in human infants and its relation to sleep, circadian cortisol activity, and behavioral distress. *Developmental Psychobiology*, **33**(4), 327–37.

Lenander-Lumikari, M., Johansson, I., Vilja, P. and Samaranayake, L. P. (1995). Newer saliva collection methods and composition: a study of two salivette kits. *Oral Disease*, **1**(2), 86–91.

Levine, A., Cohen, D. and Zadik, Z. (1994). Urinary free cortisol values in children under stress. *Journal of Pediatrics*, **125**(6 Pt 1), 853–7.

Levine, M. E., Milliron, A. N. and Duffy, L. K. (1994). Diurnal and seasonal rhythms of melatonin, cortisol and testosterone in interior Alaska. *Arctic Medical Research*, **53**(1), 25–34.

Lovallo, W. R. and Thomas, T. L. (2000). Stress hormones in psychophysiological research. In *Handbook of psychophysiology*, ed. J. T. Cacioppo, L. G. Tassinary and G. G. Berntson. Cambridge: Cambridge University Press.

Lucia, A., Diaz, B., Hoyos, J. *et al.* (2001). Hormone levels of world class cyclists during the Tour of Spain stage race. *British Journal of Sports Medicine*, **35**, 424–30.

Lundberg, U. and Forsman, L. (1980). Consistency in catecholamine and cortisol excretion patterns over experimental conditions. *Pharmacology, Biochemistry and Behavior*, **12**, 449–52.

Lundberg, U. and Frankenhaeuser, M. (1980). Pituitary-adrenal and sympathetic-adrenal correlates of distress and effort. *Journal of Psychosomatic Research*, **24**, 125–30.

Lupien, S., Lecours, A. R., Schwartz, G. *et al.* (1996). Longitudinal study of basal cortisol levels in healthy elderly subjects: evidence for subgroups. *Neurobiology of Aging*, **17**(1), 95–105.

Maes, M., Mommen, K., Hendrickx, D. *et al.* (1997). Components of biological variation, including seasonality in blood concentrations of TSH, TT3, FT4, PRL, cortisol and testosterone in healthy volunteers. *Clinical Endocrinology*, **46**, 587–98.

Marshall, R. D. and Garakani, A. (2002). Psychobiology of the acute stress response and its relationship to the psychobiology of post-traumatic stress disorder. *Psychiatria Clinica North America*, **25**(2), 385−95.

Mason, J. (1968). A review of psychoendocrine research on the pituitary corticol system. *Psychosomatic Research*, **30**, 631−43.

McEwen, B. S. (1988). Glucocorticoid receptors in the brain. *Hospital Practice*, **23**(8), 107−11, 114, 119−21.

McEwen, B. S. (2002). *The End of Stress as We Know It*. Washington, DC: Joseph Henry Press.

McEwen, B. S. and Wingfield, J. (2003). The concept of allostasis in biology and biomedicine. *Hormones and Behavior*, **43**, 2−15.

Miller, C. S., Dembo, J. B., Falace, D. A. and Kaplan, A. L. (1995). Salivary cortisol response to dental treatment of varying stress. *Oral Surgery, Oral Medicine, Oral Pathology, Oral Radiology and Endodontics*, **79**(4), 436−41.

Miller, D. and O'Callaghan, J. (2002). Neuroendocrine aspects of the response to stress. *Metabolism*, **51**, 5−10.

Munck, A. (2000). Corticosteroids and stress. In *Encyclopedia of Stress*, ed. G. Fink. San Diego, CA: Academic Press, 570−7.

Nepomnaschy, P., Welch, K., McConnell, D., Strassmann, B. and England, B. (2005). Stress and female reproductive function: a study of daily variations in cortisol, gonadotrophins, and gonadal steroids in a rural Mayan population. *American Journal of Human Biology*, **16**, 523−32.

Nicolson, N., Storms, C., Ponds, R. and Sulon, J. (1997). Salivary cortisol levels and stress reactivity in human aging. *Journals of Gerontology. Series A, Biological Sciences and Medical Sciences*, **52**(2), M68−75.

Obel, C., Hedegaard, M., Hendriksen, T. B. *et al.* (2005). Stress and salivary cortisol during pregnancy. *Psychoneuroendocrinology*, **30**, 647−56.

Ockenfels, M. C., Porter, L., Smyth, J. *et al.* (1995). Effect of chronic stress associated with unemployment on salivary cortisol: overall cortisol levels, diurnal rhythm, and acute stress reactivity. *Psychosomatic Medicine*, **57**(5), 460−7.

Pearson, R., Ungpakorn, G. and Harrison, G. A. (1995). Catecholamine and cortisol levels in Oxford college rowers. *British Journal of Sports Medicine*, **29**(3), 174−7.

Peeters, F., Nicholson, N. A. and Berkhof, J. (2003). Cortisol responses to daily events in major depressive disorder. *Psychosomatic Medicine*, **65**(5), 836−41.

Peres, M. F. P., Sanchez del Rio, M., Seabra, M. *et al.* (2001). Hypothalamic involvement in chronic migraine. *Journal of Neurology, Neurosurgery and Psychiatry*, **71**, 747−51.

Peters, M., Godaert, G., Ballieux, R. *et al.* (1998). Cardiovascular and endocrine responses to experimental stress: effects of mental effort and controllability. *Psychoneuroendocrinology*, **23**, 1−17.

Pollard, T. M. (1995). Use of cortisol as a stress marker: practical and theoretical problems. *American Journal of Human Biology*, **7**, 265−73.

Pollard, T. M., Ungpakorn, G., Harrison, G. A. and Parkes, K. R. (1996). Epinephrine and cortisol responses to work: a test of the models of Frankenhaeuser and Karasek. *Annals of Behavioral Medicine*, **18**, 229–37.

Poteliakhoff, P. V. (1981). Adrenocortical activity and some clinical findings in acute and chronic fatigue. *Journal of Psychosomatic Research*, **25**, 91–5.

Price, D. A., Close, G. C. and Fielding, B. A. (1983). Age of appearance of circadian rhythm in salivary cortisol values in infancy. *Archives of Disease in Childhood*, **58**(6), 454–6.

Prinz, P. N., Roehrs, T. A., Vitaliano, P. P., Linnoila, M. and Weitzman, E. D. (1980). Effect of alcohol on sleep and nighttime plasma growth hormone and cortisol concentrations. *Journal of Clinical Endocrinology and Metabolism*, **51**(4), 759–64.

Pruessner, J. C., Wolf, O. T., Hellhammer, D. H. *et al.* (1997). Free cortisol levels after awakening: a reliable biological marker for the assessment of adrenocortical activity. *Life Sciences*, **61**, 2539–49.

Raff, H., Raff, J. L., Duthie, E. H. *et al.* (1999). Elevated salivary cortisol in the evening in healthy elderly men and women: correlation with bone mineral density. *Journals of Gerontology. Series A, Biological Sciences and Medical Sciences*, **54**(9), M479–83.

Ramsay, D. S. and Lewis, M. (1994). Developmental change in infant cortisol and behavioral response to inoculation. *Child Development*, **65**(5), 1491–502.

Rosmalen, J. G. M., Oldehinkel, A. J., Ormel, J. *et al.* (2005). Determinants of salivary cortisol levels in 10–12 year old children: a population based study of individual differences. *Psychoneuroendocrinology*, **30**, 483–95.

Sapolsky, R. (1990). The adrenocortical axis. In *Handbook of the Biology of Aging*, ed. E. Schneider and J. Rowe. San Diego, CA: Academic Press, 330–48.

(1992). *Stress, the Aging Brain, and the Mechanisms of Neuron Death.* Cambridge, MA: MIT Press.

Sapolsky, R., Romero, L. and Munck, A. (2000). How do glucocorticoids influence stress responses? Integrating permissive, suppressive, stimulatory, and preparative actions. *Endocrine Reviews*, **21**(1), 55–89.

Schmidt-Reinwald, A., Pruessner, J. C., Hellhammer, D. H. *et al.* (1999). The cortisol response to awakening in relation to different challenge tests and a 12-hour cortisol rhythm. *Life Sciences*, **64**(18), 1653–60.

Schwartz, J. and Stone, A. (1998). Strategies for analyzing ecological momentary assessment data. *Health Psychology*, **17**, 6–16.

Sephton, S. E., Sapolsky, R. M., Kraemer, H. C. and Spiegel, D. (2000). Diurnal cortisol rhythm as a predictor of breast cancer survival. *Journal of the National Cancer Institute*, **92**(12), 994–1000.

Sherman, B., Wysham, C. and Pfohl, B. (1985). Age-related changes in the circadian rhythm of plasma cortisol in man. *Journal of Clinical Endocrinology and Metabolism*, **61**(3), 439–43.

Shiffman, S. and Stone, A. (1998). Ecological momentary assessment: a new tool for behavioral medicine research. In *Technology and Methods in Behavioral*

Medicine, ed. D. Krantz and A. Baum. New Jersey: Lawrence Erlbaum Associates.

Shirtcliff, E. A., Granger, D. A., Schwartz, E. and Curran, M. J. (2001). Use of salivary biomarkers in biobehavioral research: Cotton-based sample collection methods can interfere with salivary immunoassays. *Psychoneuroendocrinology*, **26**, 165–73.

Sippell, W. G., Becker, H., Versmold, H. T., Bidlingmaier, F. and Knorr, D. (1978). Longitudinal studies of plasma aldosterone, corticosterone, deoxycorticosterone, progesterone, 17-hydroxyprogesterone, cortisol, and cortisone determined simultaneously in mother and child at birth and during the early neonatal period. I. Spontaneous delivery. *Journal of Clinical Endocrinology and Metabolism*, **46**(6), 971–85.

Smyth, J. M., Ockenfels, M. C., Gorin, A. A. *et al.* (1997). Individual differences in the diurnal cycle of cortisol. *Psychoneuroendocrinology*, **22**(2), 89–105.

Smyth, J. M., Ockenfels, M. C., Porter, L. S. *et al.* (1998). Stressors and mood measured on a momentary basis are associated with salivary cortisol secretion. *Psychoneuroendocrinology*, **22**, 353–70.

Stewart, J. and Seeman, T. (2000). Salivary cortisol measurement. *Consensus Conference Report*, John D. and Catherine T. MacArthur Research Network on Socioeconomic Status and Health.

Stone, A., Schwartz, J., Smyth, J. *et al.* (2001). Individual differences in the diurnal cycle of salivary free cortisol: a replication of flattened cycles for some individuals. *Psychoneuroendocrinology*, **26**, 295–306.

Tout, K., de Haan, M., Campbell, E. K. and gunnar, M. R. (1998). Social behavior correlates of cortisol activity in child care: gender differences and time-of-day effects. *Child Development*, **69**(5), 1247–62.

Tremblay, M. S., Copeland, J. L. and Van Helder, W. (2004). Effect of training status and exercise mode on endogenous steroid hormones in men. *Journal of Applied Physiology*, **96**(2), 531–9.

Van Cauter, E., Plat, L., Leproult, R. and Copinschi, G. (1998). Alterations of circadian rhythmicity and sleep in aging: endocrine consequences. *Hormone Research*, **49**(3–4), 147–52.

van Eck, M., Berkhof, H., Nicolson, N. and Sulon, J. (1996). The effects of perceived stress, traits, mood states, and stressful daily events on salivary cortisol. *Psychosomatic Medicine*, **58**(5), 447–58.

Walker, B. R., Best, R., Noon, J. P., Watt, G. C. M. and Webb, D. J. (1997). Seasonal variation in glucocorticoid activity in healthy men. *Journal of Clinical Endocrinology and Metabolism*, **82**, 4015–19.

Watamura, S. E., Donzella, B., Alwin, J. and Gunnar, M. R. (2003). Morning-to-afternoon increases in cortisol concentrations for infants and toddlers at child care: age differences and behavioral correlates. *Child Development*, **74**(4), 1006–20.

Watamura, S. E., Sebanc, A. M. and Gunnar, M. R. (2002). Rising cortisol at childcare: relations with nap, rest, and temperament. *Developmental Psychobiology*, **40**(1), 33–42.

Weitzman, E. D., Fukushima, D., Nogeire, C. *et al.* (1971). Twenty-four hour pattern of the episodic secretion of cortisol in normal subjects. *Journal of Clinical Endocrinology and Metabolism*, **33**(1), 14–22.

White, B. P., Gunnar, M. R., Larson, M. C., Donzella, B. and Barr, R. G. (2000). Behavioral and physiological responsivity, sleep, and patterns of daily cortisol production in infants with and without colic. *Child Development*, **71**(4), 862–77.

Wittersheim, G., Brandenberger, G. and Follenius, M. (1985). Mental task-induced strain and its after-effect assessed through variations in plasma cortisol levels. *Biological Psychology*, **21**(2), 123–32.

Wüst, S., Federenko, I., Hellhammer, D. H. and Kirschbaum, C. (2000). Genetic factors, perceived chronic stress, and the free cortisol response to awakening. *Psychoneuroendocrinology*, **25**(7), 707–20.

Wüst, S., Kirschbaum, C. and Hellhammer, D. (1992). Smoking increases salivary cortisol. In *Assessment of Hormones and Drugs in Saliva in Biobehavioral Research*, ed. C. Kirschbaum, G. Read and D. Hellhammer. Seattle, WA: Hogrefe and Huber, 51–7.

Zeier, H. (1994). Workload and psychophysiological stress reactions in air traffic controllers. *Ergonomics*, **37**(3), 525–39.

Zeier, H., Brauchli, P. and Joller-Jemelka, H. I. (1996). Effects of work demands on immunoglobulin A and cortisol in air traffic controllers. *Biological Psychology*, **42**(3), 413–23.

6   *Measuring physiological changes in the cardiovascular system: ambulatory blood pressure*

GARY D. JAMES

Introduction

Blood pressure is the most familiar and probably the most difficult to interpret measure in anthropological studies of stress and human variation. The difficulty in interpretation arises from the use of single blood pressure measurements as an indicator of stress. The dogma that each person has specific blood pressure numbers that can be used to ascertain pathology (see for example, JNCVII, 2003) has directed the attention of anthropological researchers to the between-individual distribution of these numbers as a means of studying the evolutionary and health effects of stress on blood pressure, and away from the enormous within-individual variation that actually characterizes the adaptive value of blood pressure as it responds to the dynamic stressors of everyday living (James, 1991; Pickering, 1991).

Some 35 years ago, ambulatory monitors were developed which could capture intra-individual diurnal variation in blood pressure (James, 1991). Hundreds of subsequent studies that have employed ambulatory monitoring have shown that a myriad of external environmental stressors, cognitive processes and behavior contribute substantially to the intra-individual diurnal variation in blood pressure. It is this variation that allows people to adapt to the constantly changing challenges that define their everyday life. It is also this variation that is an important contributor to cardiovascular morbidity and mortality. The purpose of this chapter is to outline the issues, methods, and techniques that are related to the study of diurnal blood pressure variation as it occurs in response to the tribulations of everyday life using ambulatory blood pressure monitors.

158

What is blood pressure and how is it commonly measured?

Blood pressure is a dynamic, physiological consequence of the biophysics of blood circulation. It is a measure of the force of pulsatile blood flow in the arterial system. As the muscles of the heart contract and relax in a rhythmic fashion, blood is ejected from the left ventricle into the aorta, propagating a pulse wave of blood that extends throughout the cardio-vascular system (Blank, 1987). Systolic pressure is an estimate of the maximum pressure (that exerted against the arterial wall at the peak of ventricular contraction) and diastolic pressure is an estimate of the minimum pressure (that exerted when the heart is at rest; just prior to the next heart contraction) of the pulse wave (Blank, 1987; O'Rourke; 1990; James and Pecker, 1994) (Figure 6.1). As every heart contraction initiates a pulse wave, a person whose heart averages 90 beats per minute will generate 129 600 pulse waves over the course of one twenty-four hour period; therefore that same person will also have 129 600 systolic and diastolic blood pressures during the period as well (James *et al.*, 1988; James, 1991; James and Pecker, 1994; James and Brown, 1997; James, 2001).

As a pulse wave propagates through the cardiovascular system to smaller and smaller branching arteries, increasing diffusion of the blood flow changes its shape and hence the measured blood pressure (O'Rourke, 1990). Thus, by convention, for clinical, medical and most research purposes, the maximum and minimum of the pulse wave (systolic and diastolic blood pressure) are estimated from the blood flow through the brachial artery, usually in the non-dominant arm

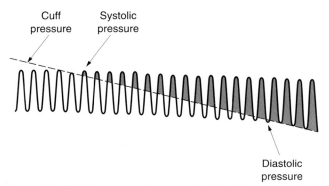

Figure 6.1. Blood pressure pulse wave, showing the measure of systolic and diastolic pressure (modified from James and Pecker, 1994).

(i.e. left arm for right dominant people). The values are determined in a process where blood flow is first occluded and then re-established in the artery using an air-inflated bladder inside a cuff, which is attached to a mercury column or gauge. A listener, using a stethoscope placed just distal to the arterial occlusion, slowly releases the pressure inside the bladder (at a rate of approximately 2 mmHg per second), recording systolic pressure as the value registered on the column or gauge when sound is first heard (Korotkoff phase I) and diastolic pressure when all sound disappears (Korotkoff phase V) (James, 1991; Pickering and Blank, 1995). At the above rate of bladder deflation, it will take about 20 to 30 seconds to record the appearance and disappearance of sound, in which time there will usually be between 30 and 50 heart contractions. Since there is both a systolic and diastolic pressure for each contraction, the systolic pressure (recorded when sound is first heard) is from a different pulse wave than the diastolic pressure (when sound disappears). Thus, the estimates of systolic and diastolic pressure as measured by this auscultatory method, do not directly correspond (O'Rourke, 1990; James and Pecker, 1994; Pickering and Blank, 1995). Rather, a single blood-pressure measurement, as taken by the method just described, represents independent point estimates of the maximum and minimum force of the continuous, pulsatile blood flow in the brachial artery.

Anything that will affect the strength of the contraction of the left ventricle of the heart will, by definition, alter the amplitude and frequency of the pulse waves (and hence the systolic and diastolic pressure). The total amount of fluid volume in the arterial system, the diameter of the arterial lumen, and arterial elasticity will all also affect the level of blood pressure (Rowell, 1986; James, 1991; Laragh and Brenner, 1995). Change in any of these structures or the endocrine, organ, or humoral factors that regulate them will have further effects on the pulse waves. For example, shifts in the levels of circulating vasoactive hormones that occur as part of the stress response (such as increases in circulating epinephrine, norepinephrine and angiotensin II) will directly influence blood pressure (James, 1991; James and Brown, 1997).

Finally, the pulse waves are not generated in a regular harmonic fashion (each with identical amplitude, frequency and base). Systolic and diastolic pressures change continuously to maintain an adequate flow of blood to the various body organs and tissues so that homeostasis at the tissue level is maintained (James and Pickering, 1993). The change in blood pressure can be dramatic and very quick; 10% or more within seconds, depending upon the oxygen needs of the tissues (Blank *et al.*, 1991; James and Pecker, 1994; Pickering and Blank, 1995). The pressures exerted at the pulse maxima and minima over the course of one 24-hour period can easily vary up to 100 mmHg depending upon the activity and adaptive needs of the individual (James, 1991; Pickering, 1991; James, 2001). Blood pressure, therefore, is an inherently unstable and variable function, and a single blood pressure measurement is nothing more than a point estimate of a perpetually moving target. The constant change in pressure with physiological need identifies blood pressure as a stereotypic example of allostasis — the physiological process of maintaining stability through change (Sterling and Eyer, 1988; Sterling, 2004).

In sum, blood pressure is an estimate of the maximum and minimum force of continuous pulsatile blood flow produced by rhythmic heart contractions, and is modified by several endocrine and physiologic factors. It is very easy to take a casual blood pressure measurement with a stethoscope and an adjustable air-inflatable bladder. However, this single determination does not tell the researcher anything about the continuing adaptation of blood pressure to changing conditions. Rather, any given measurement only estimates the amplitude of the pulse wave of blood in the brachial artery under the given circumstances when it is taken (James, 1991).

The unreliability of standardized auscultatory blood pressure measurements

Over the past 60 to 70 years, many researchers and clinicians have documented that a variety of psychological, situational and social factors can substantially affect the measured value of single blood pressure measurements taken under standardized conditions (see Pickering [1991]; James [1991]; O'Brien *et al.* [2001] and White [2001] for relevant discussions). These include but are not limited to the emotional and psychological state of the subject, familiarity between the measurer and the subject, gender or ethnic difference between the measurer and subject, the temperature in the room, and the social status of the measurer (physician vs. nurse or technician). Some of these factors have been suggested to contribute to "white coat hypertension," a phenomenon whereby a person has blood pressure measured over 140/90 (the hypertension Rubicon) in the standardized conditions of the doctor's office but "normal" blood pressures outside that venue (Pickering *et al.*, 1988). Somewhere between 20 and 40% of the people diagnosed with hypertension from clinical measurements have "white coat hypertension" (O'Brien *et al.*, 2001). The "white coat" effect can persist for the lifetime of the person, or may diminish with time (White, 2001). Because the effect increases the auscultatory blood pressure measurements of only a proportion of the population and thus alters the "true" distribution of between-individual clinical seated blood pressures (the blood pressure numbers used to assess pathology as noted above), it is quite possible that relationships reported between various stress-related characteristics, social constructs and psychological characteristics and single blood pressure measurements may be biased. Ambulatory blood pressure measurements avoid the "white coat" pitfalls of standardized clinic measurements because there are no human listeners who can influence the measurement, and the measurements are taken in the familiar surroundings of real life, not in the contrived context of the clinic (Pickering, 1991; White, 2001).

Why is blood pressure considered a physiological indicator of stress?

The interest in blood pressure as a stress marker stems from the fact that blood pressure elevation is a primary outcome in the neuroendocrine cascade of the "Fight or Flight" response (Cannon, 1915) and "General Adaptation Syndrome" (Selye, 1956) (see Chapters 4 and 5 of this book). For example, epinephrine and norepinephrine (catecholamines) excite various types of alpha and beta receptors located in the endothelial musculature and heart (see Chapter 4). When a change in psychological state or environmental conditions initiates a response that results in an increase in the secretion of the catecholamines, the receptors are either excited or down regulated which then, in turn, alter the vasculature or heart contractility leading to an increase in blood pressure. It should be realized that many other internal and external factors can and do influence blood pressure variation as well (Laragh and Brenner, 1995). However, there are many studies that have linked elevation of catecholamines with elevated blood pressure or blood pressure variability, demonstrating this stress-related link, both in the laboratory and in real life (Franklin *et al.*, 1986; James *et al.*, 1993; James and Brown, 1997; Portaluppi and Smolensky, 2001; Brown *et al.*, 2003).

Why are stress-related changes in blood pressure important to study?

There is accumulating evidence to suggest that diurnal stress-related cardiovascular changes, as reflected in the daily variation and level of ambulatory blood pressure, may contribute to the rate of cardiovascular disease and death, and are in fact a better predictor of cardiovascular morbidity and mortality than clinic measurements (Perloff *et al.*, 1989; Pickering and James, 1994; Verdeccia *et al.*, 1995; Verdeccia and Schillaci, 2001). Specifically, elevated daytime ambulatory pressure (Perloff *et al.*, 1989; Verdeccia and Schillaci, 2001), elevated daytime ambulatory pressure variation (Pickering and James, 1994) and a lack of noctural decline in ambulatory blood pressure (Verdeccia *et al.*, 1995; Verdeccia and Schillaci, 2001) have all been shown to predict cardiovascular morbidity and mortality better than blood pressures measured in a clinical setting. It is likely that the daily stress from contemporary lifestyles significantly contributes to all of these ambulatory blood pressure parameters (higher daytime pressure, greater daytime pressure variability and a lack of nocturnal decline). Because of the relationship between stress-related blood pressure changes and cardiovascular outcomes, stress-related blood pressure change could be seen as a marker for the natural selective process associated with the psychosocial and sociocultural stressors that define contemporary lifestyles.

Ambulatory blood pressure monitoring: technical aspects

Since their initial development in the late 1970s (Pickering, 1991), automatic ambulatory blood pressure monitors have evolved technologically. They have become smaller and lighter (much less cumbersome), are much more accurate in their measurement ability, and, as with all technological advancement, are significantly cheaper to use. There is now a burgeoning medical market for ambulatory blood pressure monitors and there are several manufacturers. Table 6.1 lists selected devices that are currently in clinical and research use, with company contact numbers and website information. Many of these devices measure pressure in the same manner as a listener with a stethoscope. That is, the device inflates a bladder, which occludes the brachial artery. At a point when the device detects no sound (when the artery is completely occluded) the device stops pumping air into the bladder and begins releasing the pressure at a constant rate. A microphone attached to the cuff that is placed or taped

Table 6.1. *Selected ambulatory blood pressure monitors and manufacturers*

Monitor	Contact (phone)	Website
Spacelabs 90207/90217	1-800-522-7025	www.spacelabs.com
Welch Allyn 6100-200	1-800-535-6663	www.welchallyn.com
SunTech Medical OSCAR 2	1-800-421-8626	www.suntechmed.com
Koven Technology, NISSEI DS-250	1-800-521-8342	www.koven.com
A & D Medical, TM-2430	1-888-726-9966	www.lifesourceonline.com

over the brachial artery detects the appearance and disappearance of Korotkoff sounds (Korotkoff Phases I and V). Usually, about 10 mmHg below the point when the last sound is detected, the monitor rapidly releases the remaining pressure in the bladder, records the measurement and then prepares for the next pressure measurement cycle.

Another common method of blood pressure detection, the oscillometric technique, is also widely used by ambulatory monitors (James, 1991; Pickering and Blank, 1995). In this technique, systolic and diastolic pressures are calculated from a reflected oscillating pulse wave that is generated inside the deflating bladder as blood flow is slowly established in the occluded brachial artery. The maximum oscillation of this pulse wave is thought to correspond to mean arterial pressure. Because there is a mathematical relationship between mean arterial pressure and systolic and diastolic pressure (i.e. 1/3 systolic + 2/3 diastolic pressure = mean arterial pressure), systolic and diastolic pressure can be calculated once mean arterial pressure is known. The actual calculation is done via a proprietary company algorithm that differs with each manufacturer and is stored in the computer memory of the monitor. Since every arm is different in terms of size, shape, tissue depth and bone mass (all factors which influence the ability of the bladder to occlude the artery) (Harshfield *et al.*, 1989), the shape of the oscillometric envelopes for each individual varies. The algorithms take account of this shape variation in formulating the estimates of systolic and diastolic pressure.

Operationally, the monitors have an air pump that inflates the artery-occluding bladder at pre-programmed intervals. These intervals can be set to as short as every five minutes, but can be as long as every hour or two. There is no standard for the interval between measurements, and for some purposes, a researcher might want pressures to be taken at random intervals (which many of the monitors can now do). Many researchers set the monitors for multiple fixed cycles, shorter during the day (every 15 minutes, say from 8 a.m. to 10 p.m.) and longer during sleep (every 30 minutes, from 10 p.m. to 8 a.m.). The pre-programming is done with manufacturer provided software (see below). Finally, two to four AA or AAA batteries power the air pumps of the current generation of monitors, while a back-up lithium battery powers computer memory storage.

How has the accuracy of the devices been established?

How close are the measurements taken by the monitors to the actual blood pressure in the artery? This question is related to the much broader question of how accurate auscultatory blood pressure measurements are, in general. Simultaneous blood pressure measurements taken by two listeners using a double-headed stethoscope are considered to be equivalent if they are within ± 5 mmHg of each other (Pickering, 1991; Pickering and Blank, 1995). Likewise, simultaneous measurements taken by an automatic device such as an ambulatory monitor and a listener are considered to correspond if they are within ± 5 mmHg of each other (AAMI, 1992; White *et al.*, 1993). What this means is that the "standard" to which all devices are manufactured is the "physician's ears."

The devices must maintain their accuracy over a broad range of pressures, and the testing protocols, such as that of the Association for the Advancement of Medical Instrumentation (AAMI, 1992; White *et al.*, 1993) and the British Hypertension Society (BHS) (O'Brien and O' Malley, 1990, O'Brien *et al.*, 1993), are quite rigorous. For example, the AAMI standards specify that there must be 85 subjects evaluated and this sample should be as heterogeneous as possible, providing a wide range of diastolic and systolic pressures as well as arm sizes (White *et al.*, 1994). In addition, there must be two trained listeners making simultaneous and blinded blood pressure determinations along with the device for every subject, with three sets of comparative blood pressure measurements obtained for each subject in the supine, seated and standing positions (White *et al.*, 1993; 1994). Almost all the devices currently on the market made by major manufacturers (see Table 6.1) meet or exceed the AAMI and BHS standards, with accuracy assessed over a range of systolic pressures of about 80−220 mmHg and a range of diastolic pressures of about 50−130 mmHg.

Finally, there are other subject-related factors than can compromise the accuracy of the pressures measured by ambulatory monitors, but these factors also affect standard auscultated blood pressure measurements as well. For example, among the elderly, the ability to make accurate blood pressure measurements using a cuff (bladder) occlusion method may be compromised by physical factors associated with the arterial structure, which can result in an overestimation of blood pressure. Sclerotic or calcified (non-elastic) arteries (which are more common in the elderly) may require a pressure greater than the actual pressure produced by the blood pulse waves to fully occlude the arterial lumen. Thus, the pressure recorded will be higher than the actual pressure. In addition, tissue atrophy in the arm among the elderly can also cause difficulty in making the measurements. That is, there is a reduction of subcutaneous tissues such as fat and skeletal muscle in the elderly that maintains the structural integrity of the arm (Harshfield *et al.*, 1989). Blood vessels, including the brachial artery, can become more mobile within the arm without these supporting tissues. This morphology can cause inaccuracy of blood pressure measurements because the altered shape and decreased elasticity of the tissues compromises the ability of the bladder to properly occlude the artery. The mobility of the artery may also affect the ability to make a measurement in some monitors because the microphone that detects Korotkoff sounds may not remain directly above the artery during bladder deflation, even if the microphone is taped to the skin. What this means is that to use ambulatory monitors in the assessment of stress, the subjects in the study need to have limited arterial injury (arteriosclerosis or arterial calcification) and they must also have a structurally sound arm.

The process of collecting ambulatory blood pressure data

To begin the process of collecting ambulatory blood pressure data, the monitor is first initialized via a computer, using manufacturer-provided software. Initialization is the process whereby the parameters by which the monitor will take blood pressure readings are set. The initialization software is similar across the various manufacturers. The software is "menu driven," and all use nested drop-down menus to facilitate the interface with the monitors. The manufacturers provide manuals that describe in detail how to go through the menus. The initialization parameters include defining subject-identifying information, setting the frequency of measurement, setting the method of taking measurements (Korotkoff sound or oscillometric), and setting whether or not the measurements will be displayed as they are taken. Practically, what this means is that a researcher will also need a laptop computer to do ambulatory blood pressure monitoring in the field.

After initialization, the monitor should be calibrated to a mercury column or gauge by taking simultaneous auscultatory measurements on the study subject. This is accomplished by hooking the monitor, blood pressure cuff (bladder) and mercury column together using a "Y" piece that is usually provided by the manufacturer for the purpose. The reason for calibrating the monitor is to evaluate whether it is taking "accurate" pressures, i.e. taking measurements that are within $\pm$ 5 mmHg of a simultaneous listener. If the monitor does not calibrate, reposition the cuff and continue the calibration process. At least three successive simultaneous (calibration) readings should agree to within 5 mmHg. Once the monitor is calibrated, it can be put in a carrying pouch and given to the subject (Figure 6.2). The pouch can be worn on a belt or a halter (pouch and halter are provided by the manufacturer).

In order for the monitor to take valid readings during the day, the subject must keep their arm still for the entire inflation and deflation cycle of the cuff (bladder), which is usually about 45 seconds. That means that if they are very physically active, they must stop whatever they are doing when the inflation cycle starts. To facilitate the cessation of activity, monitors can be set to "beep" five seconds prior to initiating a reading when it is initialized. If the subject is too active so that

Figure 6.2. Spacelabs 90207 ambulatory blood pressure monitor.

a reading cannot be taken, the monitors will register an error code for that reading, and will make a second measurement attempt two minutes after the error. If that reading fails, the monitor will revert to the set measurement cycle, so that the next reading would occur fifteen minutes later (if that was the set cycle).

You do not want the monitor to "beep" when it takes a reading during the sleep period of the subject, as this could be disruptive. The monitor software provides a means of setting "no beep" when the frequency of measurements during the sleep period is set.

The monitors are not infallible in that they will record invalid blood pressures. The monitors can be set to disregard systolic or diastolic measurements that are too high or too low such as diastolic measurements over 150 or under 40. However, there are pressures such as 100/95 or 120/115 that would meet the individual criteria of systolic and diastolic acceptance, but seen as a pair they indicate a physiologically impossible state. Specifically, high systolic and diastolic pressures with a very narrow pulse pressure (difference between systolic and diastolic pressure) would indicate a high pressure system that delivers virtually no blood flow. Formulas have been developed which take account of the bivariate nature of systolic and diastolic pressure, that provide guidance as to what an appropriate pulse pressure would be for given levels of systolic and diastolic pressure (see for example Pickering *et al.*, 1982; Clark *et al.*, 1987, 1989). Using these criteria, invalid pressures can be edited out of the data.

Making sense of ambulatory blood pressure variation: direct observation and diaries

From the standpoint of evaluating how everyday stressors affect ambulatory blood pressure measurements, the most critical part of the monitoring procedure is the collection of data that defines the environmental conditions and subject habitus during the day of the monitoring. This data forms the basis of the information that indicates why the measurements vary.

There are two ways to determine what the specific conditions are during each individual ambulatory blood pressure measurement. The first is through direct observation. Specifically, a person other than the study subject continuously watches and records, either in a journal or electronically, the extant conditions when the monitor goes off. This approach is necessary when small children, demented elderly or illiterate people are being evaluated (Ice *et al.*, 2003), but it is really only practical when subjects are institutionalized. A drawback of direct observational assessments is that personal data, such as psychological state or mood cannot be collected directly.

The second method involves having subjects self-report the ambient conditions in a diary. Specifically, subjects write down or input into a hand-held computer the various parameters that are manifest when the blood pressure cuff inflates, including both personal and environmental factors. When the monitor is applied in the morning, time should be spent explaining the diary procedure, detailing what the subject needs to do when the monitor takes a reading. The general instruction is that immediately after the monitor takes a reading, the subject should record the conditions that were extant when the reading occurred. This assessment could include a variety of psychological evaluations as well as the ambient environmental, physiological and social conditions. What information is relevant and should be collected in the diary? This will vary for each study depending upon what hypotheses are being tested. However, at minimum, posture, location, activity and perhaps an assessment of mood should be included. The manufacturers of the monitors will include a diary (filled out with a pencil) that has space for these relevant reports. More sophisticated diaries with a great deal of additional psychological and social information have been developed that can be used for example, with a Palm Pilot®, so that the diary can directly interface with a computer (Kamarck *et al.*, 1998) (see Chapter 3). From a practical standpoint, the type of diary used is more likely to be dictated by the sophistication of the subjects in the study rather than by the nature of the research question. In populations that are not technologically savvy, simple pencil and paper diaries are more appropriate. Figure 6.3 shows a sample paper and pencil diary. It should also be noted that if a more complex diary is going to be used, the interval between measurements should probably be increased. The reason being that if the diary takes five or ten minutes to fill out, pressures taken every 15 minutes may only reflect the stressful effects of filling out the diary, and not the effects being evaluated by the diary.

If the diary data are collected by pencil and paper, then it must be entered into the computer in order to evaluate its association with the blood pressures taken by the monitor. Diary information can be added to the pressures using the manufacturer-provided software. The measurements can also be exported out of the manufacturer software and into a spreadsheet program such as Excel in Microsoft Office®. Once in this

NAME _____

Location codes	Position codes	Mood codes
H = Home	S = Sitting	Place a number from 1 (low) to
W = Work	U = Upright/Standing	10 (high) in the appropriate box
M = Miscellaneous	R = Reclining	(Skip if mood is neutral)

TIME	LOCATION	POSITION	MOOD				ACTIVITIES
			HAPPY	SAD	ANGRY	ANXIOUS	

Figure 6.3. Ambulatory blood pressure diary (modified from James, 2001).

spreadsheet format, the data can be analyzed using SPSS, SAS or other statistical packages.

Finally, how important is it that the diary entries occur at the exact time when each blood pressure measurement takes place? Might filling in diary entries after the fact affect the accuracy of reporting? Stone *et al.* (2003), in a study of pain reporting, found that 94% of patients were compliant in writing down pain assessments when there was a signal to do so. As previously noted, with blood pressure monitoring the monitor signals five seconds before the measurement; hence it is likely that the patient will usually record something in their diary at the time of measurement. Conditions such as posture, place of measurement and activity can probably be added at a debriefing when the subject returns the monitor, without sacrificing the accuracy about those conditions at each measurement. However, mood, mood intensity and other psychological assessments will be likely to be less accurate if filled in later (Stone *et al.*, 2003).

Study design

As with all research, in conducting a study of ambulatory blood pressures and stress it is best to proceed with a design or plan in mind, one that will guide not only what data will be collected, but also one that is tied to the analytic technique (statistics) that will be used to evaluate or test the study hypotheses. There are two general approaches to studying stress using ambulatory blood pressures. The first employs what might be termed a "natural experimental" approach in which there are a priori design elements, which facilitate direct comparisons. This approach has its roots in the study of stress in the laboratory, where blood pressure reactivity to various "stressful" tasks is evaluated (see for example, Pickering and Gerin, 1990; Kamarck *et al.*, 2003; Linden *et al.*, 2003). In the laboratory, a baseline condition is established and then the paricipant undertakes various "stressful" tasks. The difference between the baseline and task measurements determines the magnitude of the stress response. Laboratory experimental designs are balanced in that there are a set number of measurements during each condition and all participants experience the baseline and all stressful tasks. Moving this paradigm to a "natural experimental" setting requires some modification, in that no true baseline can be established. However, as I have noted elsewhere (James, 1991) the environments people live in are often heterogeneous, and there are clear transitions between the microenvironments that characterize this variability, which can be studied to assess the stress of everyday living. For example, a suburban commuter may have a structured, urban work microenvironment where economic tasks are performed, where social interaction occurs with non-related individuals and where a specific occupational hierarchy dictates social relationships. The parameters of this microenvironment contrast sharply with the suburban home microenvironment, where domestic tasks and leisure activity occur in a social context where interactions are with relatives and neighbors. The allostatic variation in blood pressure required to adapt to the microenvironmental changes between these work and home

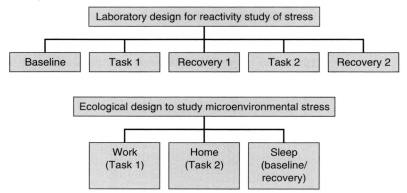

Figure 6.4.   Comparison of laboratory blood pressure reactivity study design with natural experimental design for assessing ambulatory blood pressure variation.

microenvironments can be evaluated as a "natural experiment" of the blood pressure response to the stress of everyday living. A microenvironment that is similar for all subjects (such as overnight sleep, or more specifically, lying quietly in a dark room) can act as a pseudo-baseline in this "natural experiment." This type of approach has features of a laboratory study of stress reactivity in that all subjects experience all microenvironments and have similar numbers of blood pressure measurements within each microenvironment (Figure 6.4). The variation in average blood pressures across these microenvironments can be seen as the reactive blood pressure response to the stresses inherent in each of the microenvironments.

The second approach is one where no prior design is employed. Each blood pressure taken on each study subject is considered a unit of measurement (often called ecological momentary data) and is evaluated using *post hoc* or a posteriori statistical models (see for example Schwartz *et al.*, 1994; Kamarck *et al.*, 1998; Brondolo *et al.*, 1999; Gump *et al.*, 2001). The diary data is used to define factors that contribute to the variation in the individual blood pressure measurements. In this approach, the choice of diary reporting alternatives is critical; that is, the factors chosen to have reported in the diary and how the diary factors are recorded will dictate how blood pressure variation is evaluated. In addition, care must be taken in specifying the statistical model that will be used for analysis because some factors of interest will be relevant for each individual blood pressure measurement (such as posture or mood) but other factors will be relevant to the person on whom the blood pressures were taken (such as age and BMI). The statistical models needed to evaluate this ecological momentary data are quite sophisticated as there is also imbalance in the number and types of factors experienced by each study subject (some subjects may experience anger but others may not) and in the number of experiences of a factor across subjects (one individual may have five blood pressures taken when they are angry, while another subject may have only one). These models are discussed in greater detail in Chapter 9. In the a posteriori approach, the sources of blood pressure variation are identified, and their contribution to either increasing or decreasing the values of individual blood pressure measurements relative to some standard value is estimated (James, 2001).

Analysis of ecological momentary blood pressure data in studies that have used an a posteriori approach has been undertaken using raw (Schwartz et al., 1994; Brondolo et al., 1999; Gump *et al.*, 2001; Kamarck *et al.*, 2003) and standardized (James *et al.*, 1986; Brown *et al.*, 1998; Ice *et al.*, 2003) data. Studies that have used raw data with this approach have used a model with this general form:

$$BP = overall\ mean + person + location + posture + activity$$
$$+ mood + \cdots + error$$

The person effect estimates the between-person variation in pressure and the other factors are examined for their contribution to the within-person variation in pressure. All estimated effect sizes are relative to the overall mean. Using standardized data (z-scores standardized on individual subject means, so that every subject has a mean of 0 and a variance of 1; and thus the overall population mean is also 0), the model can be simplified to:

$$BP = location + posture + activity + mood + \cdots + error$$

The estimated within-person effect sizes are relative to the person's mean (0) and will vary from person to person based on the person's standard deviation (square root of the person's variance). Even though the person effect in this model is 0, for analytic purposes, person as a factor contributing to blood pressure variance should probably still be included in the model, in order to properly estimate the true overall variance (see Chapter 9). However, the estimated variance will probably still be a reasonable estimate (although slightly biased) if the person effect is not included, provided the sample size is large. Interestingly, to date, no one has yet tried to evaluate whether the use of raw or standardized data is equivalent, in terms of evaluating within-person effect sizes.

Regardless of approach, variation in several within-person factors has consistently been found to contribute to ambulatory blood pressure variation, although the exact sizes of the effects vary from study to study. These include postural variation, variation in microenvironment, variation in mood and variation in activity (see James, 1991; James and Brown, 1997 and James, 2001 for discussion).

Finally, in evaluating the results of ambulatory blood pressure studies (whether using a natural experimental or *post hoc* designs), how meaningful are the estimated effect sizes or differences? That is, how big do effects or differences have to be before one can say they are biologically meaningful? As previously noted, any given measurement is accurate to within 5 mmHg to the "physician's ears," which themselves may be biased. Given that fact, one might argue that any effect that is smaller than 5 mmHg may simply be "noise" even though it is statistically significant.

Evaluating the ambulatory blood pressure literature

Before embarking on an evaluation of the ambulatory blood pressure literature, it is important to emphasize that the results of studies of "stress" and ambulatory blood pressure measurements and "stress" and single blood pressure measurements are not comparable and provide very different information regarding the associations between stressors and physiological outcome. The focus of ambulatory blood pressure studies is on those factors that contribute to intra-individual blood pressure change during the processes of everyday living. The conditions under which the measurements are taken are intentionally allowed to vary in order to determine how the individual adapts. This approach is in sharp contrast of studies evaluating single blood pressure measurements per subject, which focus on factors that contribute to between-individual differences in measurements taken in a standard circumstance. The environment in which measurements are taken is intentionally held constant. Systolic and diastolic blood pressure values are analyzed as relatively immutable quantitative traits, like adult stature. In studies of ambulatory blood pressure, factors such as anger or microenvironment (being at work or home) are examined to determine if they change an individual's blood pressure. In studies that examine single blood pressures, the same factors are examined to determine why two individuals have different measurements. The factors evaluated in ambulatory studies are treated as varying "states" confronted by the individual, while those in single measurement studies are treated as immutable "traits" or situations of the individual.

The majority of studies reporting on ambulatory blood pressure are found in the hypertension, health psychology, behavioral medicine, physiology and psychophysiology literatures. How ambulatory blood pressure monitors are used and the emphasis of the study results are quite different across these fields.

In the hypertension literature, emphasis is on the average value of all the pressures taken across the day, or the difference in values from waking to sleep (dipping). The interest in this literature is on the use of ambulatory blood pressures to better define the medical condition of hypertension, to improve the estimation of the cardiovascular morbidity and mortality risks of sustained high blood pressure and to evaluate the efficacy of pharmacological agents in sustaining blood pressure control. There is often little interest in the mix of conditions and factors that combine to produce the daily average blood pressures. Thus, this literature is of limited use in understanding blood pressure variation as a response to stress.

From the standpoint of blood pressure response to psychosocial stress, the health psychology, behavioral medicine and psychophysiology literatures are most revealing, as these studies focus on behavioral and psychological factors as they affect blood pressure variation across the day. In evaluating these literatures, however, be aware that there is variation across studies in the methods of behavioral measurement as well as statistical modeling. For example, in evaluating psychological factors, different methods will be employed in examining the same thing. Mood determination may be evaluated by self-report of a limited number of alternatives from a diary, by an index created from questionnaire responses or from identification of a direction in a circumplex model (see Jacob *et al.*, 1999). Each of these assessments may also have a different statistical approach with different assumptions about the data. The results of each study may only be relevant for the sample of subjects that were studied.

Finally, in evaluating the health psychology, behavioral medicine and psychophysiology literature, you will also find a literature of blood pressure "reactivity." This literature refers to stress-related blood pressure changes when measurements are made in the controlled setting of laboratory research. Be aware that blood pressure reactivity (measured in the laboratory) and blood pressure variation (measured in the field) are different conceptually. In laboratory reactivity studies, stress-related blood pressure change is examined with regard to controlled isolated single stressors (for example a speech task or mental arithmetic), whereas in field studies of blood pressure variation several simultaneous stressors are evaluated for their effect on blood pressure (such as mood, temperature, task or activity and social interactions). There is current and ongoing debate as to exactly how blood pressure reactivity and diurnal variation are related (see for example Pickering *et al.*, 1990; Kamarck *et al.*, 2003; Linden *et al.*, 2003). Often, however, the extent of blood pressure reactivity is unrelated to the extent of blood pressure variation (Harshfield *et al.*, 1988; Pickering *et al.*, 1990), although recently there have been statistical attempts to link the two (Kamarck *et al.*, 2003).

Conclusions

From an evolutionary perspective, it is important to know what blood pressure is when survival is threatened as well as when it is not. Blood pressure changes in an allostatic fashion to meet actual or perceived physiological need, which may or may not adapt people to their surroundings. The constant change in blood pressure is what makes it a useful stress measure. Ambulatory blood pressure monitoring can capture this variation.

Current ambulatory monitors still use a cuff inflation method to measure blood pressure, and there are several factors that can compromise the accuracy of these measurements. Nonetheless, a great deal has been learned about the causes and consequences of high blood pressure and cardiovascular disease using ambulatory blood pressure monitors, and as statistical designs and analytic techniques improve, the value of ambulatory blood pressure monitoring in evolutionary and stress research is likely to grow.

References

AAMI (1992). *Association for the Advancement of Medical Instrumentation: American National Standards for Electronic or Augmented Sphygmoman-ometers*, 2nd edn, Arlington, VA.

Blank, S. G. (1987). The Korotkoff signal and its relationship to the arterial pressure pulse. Unpublished Ph.D. thesis, Department of Physiology and Biophysics, Cornell University Medical College.

Blank, S. G., West, J. E., Muller, F. B. *et al.* (1991). Characterization of auscultatory gaps with wideband external pulse recording. *Hypertension*, **17**, 225–33.

Brondolo, E., Karlin, W., Alexander, K., Bubrow, A. and Schwartz, J. (1999). Workday communication and ambulatory blood pressure: implications for the reactivity hypothesis. *Psychophysiology*, **36**, 86–94.

Brown, D. E., James, G. D. and Nordloh, L. (1998). Comparison of factors affecting daily variation of blood pressure in Filipino-American and Caucasian nurses in Hawaii. *American Journal of Physical Anthropology*, **106**, 373–83.

Brown, D. E., James, G. D., Nordloh, L. and Jones, A. A. (2003). Job strain and physiological stress responses in nurses and nurse's aides: predictors of daily blood pressure variability. *Blood Pressure Monitoring*, **8**, 237–42.

Cannon, W. B. (1915). *Bodily Changes in Pain, Hunger, Fear and Rage: An Account of Recent Researches into the Functions of Emotional Excitement.* New York: Appleton.

Clark, L. A., Denby, L. and Pregibon, D. (1989). Data analysis of ambulatory blood pressure readings before p values. In *Handbook of Research Methods in Cardiovascular Behavioral Medicine*, ed. N. Schneiderman, S. M. Weiss and P. Kaufman. New York: Plenum Press, pp. 317–32.

Clark, L. A., Denby, L., Pregibon, D. *et al.* (1987). A data-based method for bivariate outlier detection: application to automatic blood pressure recording devices. *Psychophysiology*, **24**, 119–25.

Franklin, S. S., Sowers, J. R. and Batzdorf, U. (1986). Relationship between arterial blood pressure and plasma norepinephrine levels in a patient with neurogenic hypertension. *American Journal of Medicine*, **81**, 1105–7.

Gump, B. B., Polk, D. E., Kamarck, T. W. and Shiffman, S. (2001). Partner interactions are associated with reduced blood pressure in the natural environment: ambulatory blood pressure monitoring evidence from a healthy, multiethnic adult sample. *Psychosomatic Medicine*, **63**, 423–33.

Harshfield, G. A., Hwang, C., Blank, S. G. and Pickering, T. G. (1989). Research techniques for ambulatory monitoring. In *Handbook of Research Methods in Cardiovascular Behavioral Medicine*, ed., N. Schneiderman, S. M. Weiss and P. Kaufman. New York: Plenum Press, pp. 293–309.

Harshfield, G. A., James, G. D., Schlussel, Y. *et al.* (1988). Do laboratory tests of blood pressure reactivity predict blood pressure variability in real life? *American Journal of Hypertension*, **1**, 168–74.

Ice, G. H., James, G. D. and Crews, D. E. (2003). Blood pressure variation in the institutionalized elderly. *Collegium Antropologicum*, **27**, 47–55.

Jacob, R. G., Thayer, J. F., Manuck, S. B. *et al.* (1999). Ambulatory blood pressure responses and the circumplex model of mood: a 4-day study. *Psychosomatic Medicine*, **61**, 319–33.

James, G. D. (1991). Blood pressure response to the daily stressors of urban environments: methodology, basic concepts and significance. *Yearbook of Physical Anthropology*, **34**, 189–210.

(2001). Evaluation of journals, diaries, and indexes of worksite and environmental stress. In *Contemporary Cardiology: Blood Pressure*

Monitoring in Cardiovascular Medicine and Therapeutics, ed. W. B. White. Totowa, NJ: Humana Press, pp. 29–44.

James, G. D. and Brown, D. E. (1997). The biological stress response and lifestyle: catecholamines and blood pressure. *Annual Review of Anthropology*, **26**, 313–35.

James, G. D. and Pecker, M. S. (1994). Aging and blood pressure, In *Biological Anthropology and Aging: An Emerging Synthesis*, ed. D. E. Crews and R. M. Garruto. New York: Oxford University Press, pp. 321–38.

James, G. D. and Pickering, G. D. (1993). The influence of behavioral factors on the daily variation of blood pressure. *American Journal of Hypertension*, **6**, 170S–74S.

James, G. D., Pickering, T. G., Yee, L. S. *et al.* (1988). The reproducibility of average ambulatory, home, and clinic pressure. *Hypertension*, **11**, 545–9.

James, G. D., Schlussel, Y. R. and Pickering, T. G. (1993). The association between daily blood pressure and catecholamine variability in normotensive working women. *Psychosomatic Medicine*, **55**, 55–60.

James, G. D., Yee, L. S., Harshfield, G. A., Blank, S. and Pickering, T. G. (1986). The influence of happiness, anger and anxiety on the blood pressure of borderline hypertensives. *Psychosomatic Medicine*, **48**(6), 502–8.

JNCVII (2003). *The Seventh Report of the Joint National Committee on Prevention, Detection, Evaluation, and Treatment of High Blood Pressure*, Bethesda, MD: NIH Publication No. 03–5233.

Kamarck, T. W., Schiffman, S. M., Smithline, L. *et al.* (1998). Effects of task strain, social conflict, on ambulatory cardiovascular activity: life consequences of recurring stress in a multiethnic adult sample. *Health Psychology*, **17**, 17–29.

Kamarck, T. W., Schwartz, J. E., Janiki, D. L., Schiffman, S. and Raynor, D. A. (2003). Correspondence between laboratory and ambulatory measures of cardiovascular reactivity: a multilevel modeling approach. *Psychophysiology*, **40**, 675–83.

Laragh, J. H. and Brenner, B. M. (1995). *Hypertension: Pathophysiology Diagnosis and Management*. New York: Raven.

Linden, W., Gerin, W. and Davidson, K. (2003). Cardiovascular reactivity: Status quo and a research agenda for the new millennium. *Psychosomatic Medicine*, **65**, 5–8.

O'Brien, E., Beevers, G. and Lip, G. Y. (2001). ABC of hypertension. Blood pressure measurement. Part III. Automated sphygmomanometry: ambulatory blood pressure measurement. *British Medical Journal*, **322**, 1110–14.

O'Brien, E. and O'Malley, K. (1990). Twenty-four-hour ambulatory blood pressure monitoring: a review of validation data. *Journal of Hypertension Supplement*, **8**(6), S11–S16.

O'Brien, E., Petrie, J., Littler, W. *et al.* (1993). An outline of the revised British Hypertension Society protocol for the evaluation of blood pressure measuring devices. *Journal of Hypertension*, **11**, 677–9.

O'Rourke, M. F. (1990). What is blood pressure? *American Journal of Hypertension*, **3**, 803–10.

Perloff, D., Sokolow, M., Cowan, R. M. and Juster, R. P. (1989). Prognostic value of ambulatory blood pressure measurements: Further analysis. *Journal of Hypertension*, **7**(Supplement 3), S3–S10.

Pickering, T. G. (1991). *Ambulatory Monitoring and Blood Pressure Variability*. London: Science

　(1995). Modern definitions and clinical expressions of hypertension. In *Hypertension: Pathophysiology, Diagnosis and Management*, ed. J. H. Laragh and B. M. Brenner. New York: Raven, pp. 17–21.

Pickering, T. G. and Blank, S. G. (1995). Blood pressure measurement and ambulatory pressure monitoring: Evaluation of available equipment. In *Hypertension: Pathophysiology, Diagnosis and Management*, ed. J. H. Laragh and B. M. Brenner. New York: Raven, pp. 1939–52.

Pickering, T. G. and Gerin, W. (1990). Cardiovascular reactivity in the laboratory and the role of behavioral factors in hypertension: a critical review. *Annals of Behavioral Medicine*, **12**, 3–16.

Pickering, T. G., Harshfield, G. A., Kleinert, H. D., Blank, S. and Laragh, J. H. (1982). Blood pressure during normal daily activities, sleep and exercise. Comparison of values in normal and hypertensive subjects. *Journal of the American Medical Association*, **247**, 992–6.

Pickering, T. G. and James, G. D. (1994). Ambulatory blood pressure and prognosis. *Journal of Hypertension*, **12**(Supplement 8), S29–S33.

Pickering, T. G., James, G. D., Boddie, C. *et al.*, (1998). How common is white coat hypertension? *Journal of the American Medical Association*, **259**, 225–8.

Portaluppi, P. and Smolensky, M. H. (2001). Circadian rhythm and environmental determinants of blood pressure regulation in normal and hypertensive conditions. In *Contemporary Cardiology: Blood Pressure Monitoring in Cardiovascular Medicine and Therapeutics*, ed. W. B. White. Totowa, NJ: Humana Press, pp. 79–138.

Rowell, L. B. (1986). *Human Circulation: Regulation During Physical Stress*. New York: Oxford University Press.

Schwartz, J. E., Warren, K. and Pickering, T. G. (1994). Mood, location and physical position as predictors of ambulatory blood pressure and heart rate: Application of a multilevel random effects model. *Annals of Behavioral Medicine*, **16**, 210–20.

Selye, H. (1956). *The Stress of Life*. New York: McGraw-Hill.

Sterling, P. (2004). Principals of allostasis: optimal design, predictive regulation, pathophysiology, and rationale therapeutics. In *Allostasis, Homeostasis and the Costs of Physiological Adaptation*, ed. J. Schulkin. New York: Cambridge University Press, pp.17–64.

Sterling, P. and Eyer, J. (1988). Allostasis: a new paradigm to explain arousal pathology. In *Handbook of Life Stress, Cognition and Health*, ed. S. Fisher and J. Reason. New York: Wiley, pp. 629–49.

Stone, A. A., Shiffman, S., Schwartz, J. E., Broderick, J. E. and Hufford, M. R. (2003). Patient compliance with paper and electronic diaries. *Control Clinical Trials*, **24**, 182–99.

Verdeccia, P. and Schillaci, G. (2001). Prognostic value of ambulatory blood pressure monitoring. In *Contemporary Cardiology: Blood Pressure Monitoring in Cardiovascular Medicine and Therapeutics*, ed. W. B. White. Totowa, NJ: Humana Press, pp. 191–218.

Verdeccia, P., Schillaci, G., Borgioni, C. *et al.* (1995). Gender, day–night blood pressure changes and left ventricular mass in essential hypertension: dippers and peakers. *American Journal of Hypertension*, **8**, 193–6.

White, W. B. (2001). *Contemporary Cardiology: Blood Pressure Monitoring in Cardiovascular Medicine and Therapeutics*. Totowa, NJ: Humana Press.

White, W. B., Berson, A. S., Robbins, C. *et al.* (1993). National standard for measurement of resting and ambulatory blood pressures with automated sphygmomanometers. *Hypertension*, **21**, 504–9.

White, W. B., Susser, W., James, G. D. *et al.* (1994). Multicenter assessment of the QuietTrak ambulatory blood pressure recorder according to the 1992 AAMI guidelines. *American Journal of Hypertension*, **7**, 509–14.

7 Measuring immune function: markers of cell-mediated immunity and inflammation in dried blood spots

THOMAS W. McDADE

Introduction

Stress is an important determinant of human immune function, but population-level research on the social ecology of immunity has lagged behind other areas of investigation. In large part this is due to methodological constraints associated with the assessment of immune function: venipuncture is a relatively invasive blood sampling procedure that requires the skills of a trained medical professional and, once collected, blood samples must be immediately assayed, or centrifuged, separated, and frozen. Obviously, these are serious impediments to field-based research, and they have hindered the exploration of stress–immunity relationships in large, representative, community-based studies.

Dried blood spots – samples of whole blood collected on filter paper following a simple finger prick – provide an alternative, minimally invasive sampling procedure. Several community-based applications have shown this to be a convenient and reliable means to facilitate sample collection, storage, and transportation, and laboratory methods have been validated for a growing number of analytes (Worthman and Stallings, 1994, 1997; Cook et al., 1998; McDade et al., 2000a; Erhardt et al., 2002; McDade and Shell-Duncan, 2002; McDade et al., 2004). Also referred to as "Guthrie papers," filter papers have been a core component of US hospital-based newborn screening programs since the 1960s, with all newborns providing a blood spot sample to screen for congenital metabolic disorders (Mei et al., 2001).

In this chapter, detailed information is provided on the collection, handling, and analysis of dried blood-spot samples for two measures of immune activity: Epstein-Barr virus antibodies and C-reactive protein. Minimally invasive methods facilitate research with diverse populations outside the confines of the clinic or the lab, and results from the first field application of these methods are shared to demonstrate their utility for future field-based research.

Stress, immune function, and disease

There is compelling evidence that stress-induced alterations in immune function are causally related to a range of disease outcomes (Kiecolt-Glaser and Glaser, 1995; Cohen *et al.*, 1998; Moynihan *et al.*, 1998; Biondi, 2001). The majority of research on the physiological pathways linking stress, immune function, and morbidity is conducted with animal models, but epidemiological as well as experimental evidence suggests analogous processes are at work in humans. Infection has been the most intensively investigated disease outcome, with stress and immunity linked to the common cold, tuberculosis, genital herpes, infectious mononucleosis, and HIV progression. Psychosocial stress has also been associated with reduced response to vaccination, a model that mimics the real-world process of pathogen exposure and immune response that is critical in defining resistance to infectious disease (Glaser *et al.*, 1992; Petrie *et al.*, 1995). Pathogen exposure alone is not sufficient to predict the occurrence or severity of infection, and psychosocial stress may in many cases be a critical cofactor (Meyer and Haggerty, 1962; Boyce *et al.*, 1977; Cohen, 1995).

Immune processes are involved in response to injury, and a series of experiments has demonstrated that the rate of wound healing is slowed by psychosocial stress. For example, in students given a standardized punch biopsy wound, healing took on average three days longer when the wound was administered prior to a major exam than it did over the summer (Marucha *et al.*, 1998). For older individuals caring for a relative with Alzheimer's disease, wound healing was delayed almost ten days compared to age-matched controls (Kiecolt-Glaser *et al.*, 1995).

Causal linkages among stress, immunity, and risk for cancer in humans are more controversial. Animal models indicate that stress-induced immunosuppression can promote metastasis and decrease survival. Results with humans are less conclusive, but suggest that stressors such as bereavement or lack of social support may contribute to disease initiation or progression (Biondi *et al.*, 1996; Biondi, 2001). Similarly, evidence is mixed for autoimmune diseases, although a number of studies have linked psychosocial factors to alterations in immunity and disease symptomatology in rheumatoid arthritis (Affleck *et al.*, 1997; Zautra *et al.*, 1997).

The association between psychosocial stress and cardiovascular disease (CVD) has been established for some time, but immune processes have only recently been implicated as potential causal mechanisms. Current research has indicated that CVD may have a substantial inflammatory component, with immunologically mediated processes centrally involved in disease initiation and progression (Tracy, 1998; Ross, 1999; Libby *et al.*, 2002). In particular, cytokines such as IL-1β, IL-6, and TNFα may contribute to the development of atherosclerotic lesions, and recent research has suggested that psychosocial stressors can upregulate production of pro-inflammatory cytokines (Black, 2002). This is a new area of research, and the potential contribution of stress-induced modulation of inflammatory processes to risk for CVD and other chronic diseases remains to be thoroughly investigated.

In sum, experimental as well as observational research highlights the relevance of stress and immune function to a range of disease processes. However, negative findings are also present, and in many cases logistical and ethical considerations have limited attempts to establish direct links between stress, immunity, and disease in humans.

Current research in psychoneuroimmunology

Psychoneuroimmunologists have employed a number of measures of immunity to investigate the effects of psychosocial stress (Schleifer *et al.*, 1986; Herbert and Cohen, 1993; Kiecolt-Glaser and Glaser, 1997). Enumerative measures include the quantification of the number and/or percentage of various white blood cells in peripheral blood (helper T cells, cytotoxic/suppressor T cells, B cells, natural killer cells, monocytes), or the quantification of immunoglobulins in blood (primarily IgG, IgM, IgA) or saliva (secretory IgA). An additional method involves the measurement of specific antibodies, such as antibodies against resident herpes viruses (Glaser *et al.*, 1985), or antibody responses following a vaccine challenge (Glaser *et al.*, 1992; Petrie *et al.*, 1995).

Commonly used functional assays of immune competence include lymphocyte proliferation and natural killer (NK) cell cytotoxic activity. To measure proliferative responsiveness (also referred to as blastogenesis), lymphocytes are incubated with mitogens that induce non-specific T and B cell division and replication. Reduced proliferation reflects down-regulated immune activity that has been associated with a range of immunodeficiency conditions (Kiecolt-Glaser and Glaser, 1997). Recently, there has been increased interest in investigating patterns of cytokine and cytokine receptor production following lymphocyte activation as additional functional measures of immunity (Marshall *et al.*, 1998; Glaser *et al.*, 2001). NK cell activity is evaluated by growing target cells (typically from a tumor cell line) in media, and then measuring the ability of NK cells to lyse the target cells. NK cell activity is thought to be important for protecting the body against damaged or altered (i.e., infected or cancerous) cells.

For the most part, these protocols require the collection of large volumes of blood through venipuncture, and prompt access to laboratory facilities for the processing and analysis of samples. This has served as a major constraint on attempts to conduct field-based, population-level research on stress and immunity, and is a primary reason why the vast majority of current work is based in laboratory or clinical settings, with relatively homogenous, opportunistic samples. A recent literature review underscores this point, stating "The typical experimental subject in psychoimmunology is a young, male, Caucasian, healthy, medical or psychology student, probably a light or nonsmoker, consuming little or no alcohol or coffee, with no history of allergy or recent infectious disease..." (Biondi, 2001, p. 202).

Despite the methodological obstacles, a number of investigators have attempted to consider a range of stressors occurring in more naturalistic circumstances. While these studies cannot attain the high degree of control possible in the laboratory, they strive for increased generalizability and external validity. Some of the earliest PNI work investigated changes in immune function surrounding bereavement, and found consistent impairments in immune function following the loss of a loved one (Irwin *et al.*, 1987). An early series of studies investigating changes in immunity during medical school examinations has also been important, with exam stress associated with impairments in multiple measures of immune function (Glaser *et al.*, 1985; Jemmott and Magloire, 1988; Glaser *et al.*, 1994). These studies also show moderating effects of social support. More recent studies have considered the immunological impact of the stress and life disruption associated with natural disasters such as earthquake or hurricane (Boyce *et al.*, 1993; Ironson *et al.*, 1997; Solomon *et al.*, 1997).

An additional series of studies explores the immunosuppressive effects of stressful personal relationships (Kiecolt-Glaser *et al.*, 1994).

Recent divorce and self-reports of poor marriage quality have been related to increased levels of herpes-virus antibodies (indicative of suppressed cell-mediated immunity), and spousal caregivers of Alzheimer's disease patients report more days of illness, have higher herpes-virus levels, heal more slowly following an experimentally induced wound, and are less likely to produce antibodies in response to a influenza vaccine than age-matched controls (Kiecolt-Glaser *et al.*, 1995, 1996). Caregivers over the age of 70 show significantly greater reductions in immune function than younger caregivers.

The evidence linking stress to human immune function is relatively strong and consistent. A meta-analysis of this literature (Herbert and Cohen, 1993) shows that stress is significantly associated with decreased numbers of T, B, and NK cells, suppressed lymphocyte proliferation and cytotoxic activity, and lower levels of secretory IgA and IgM. Antibodies to resident herpes viruses are elevated under stress, consistent with the interpretation that stress suppresses cell-mediated immune function. Based on this meta-analysis, cautious conclusions can be drawn regarding the impact of different types of stressors: 1) objective events appear to have greater effects than self-reported events, 2) long-term stressors have more consistent negative effects than acute stressors, indicating the absence of adaptation immunologically to stress, and 3) social and non-social stressors have different immunological consequences.

In addition, although this has not been a major focus of research, constitutional and/or contextual factors may influence an individual's sensitivity to stress-induced immune modulation. For example, age, personality factors, presence of an immunosuppressive disease such as AIDS, physiological reactivity to an acute stressor, levels of chronic background stress, and social support may moderate the effects of stress on immune function and disease (Boyce *et al.*, 1995; Kiecolt-Glaser and Glaser, 1995; Kiecolt-Glaser *et al.*, 1996; Uchino *et al.*, 1996; Pike *et al.*, 1997; Segerstrom *et al.*, 2001).

Prior research in PNI has demonstrated that psychosocial processes are causally related to important aspects of human immune function. The assessment of immunity is complicated by the fact that the immune system is comprised of multiple, inter-related subsystems of defense, and therefore no single measure can provide a global assessment of immunocompetence. Add to this the relative invasiveness of current methods of immunological assessment, and it is clear why the majority of research on stress and immune function has been limited to clinical or laboratory settings.

Measuring cell-mediated immunity: antibodies against the Epstein-Barr virus in dried blood spots

A practical method for assessing immunity in field-based stress research must overcome the logistical constraints associated with venipuncture without sacrificing validity or reliability. Whole blood spots collected from a simple finger prick provide a minimally invasive alternative for collecting blood, and antibodies against the Epstein-Barr virus (EBV) can be quantified as an indirect measure of cell-mediated immunity.

The EBV antibody model

The Epstein-Barr virus is a ubiquitous herpes virus to which approximately 90% of adults in industrialized nations, and nearly 100% of adults in developing nations, have been exposed (Henle and Henle, 1982). Once infected, individuals permanently harbor EBV, and adequate cell-mediated immune function is critical for maintaining the virus in a latent state. Stress-induced immunosuppression allows EBV to reactivate and release viral antigens into circulation, to which a humoral antibody response may emerge (Glaser *et al.*, 1991). As a result, levels of antibodies against EBV antigens provide an indirect measure of an aspect of cell-mediated immune function, such that increased EBV antibody titers indicate lower cell-mediated immunity (Figure 7.1).

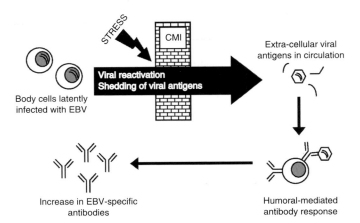

Figure 7.1. Model of the relationships among stress, cell-mediated immunity, and EBV antibody titers. An increase in EBV-specific antibodies indicates relative impairment of cell-mediated immune performance.

Although this model may at first seem counterintuitive — an *increase* in EBV antibody titer, itself an aspect of immune function, is interpreted as indicating a *decrease* in cell-mediated immune performance — it is important to recognize that cell-mediated immune processes are primarily involved in maintaining viral latency, and that humoral-mediated immunity (including the production of antibodies) represents a second line of defense that is called upon when cell-mediated processes fail to control the virus. In addition, increased EBV antibody titers have been associated with reduced EBV-specific T-cell proliferation and reduced cytotoxic T-cell lysis of EBV-infected target cells, consistent with the interpretation that EBV antibodies reflect broader cell-mediated immune performance (Glaser *et al.*, 1987; Glaser *et al.*, 1993).

The utility of the EBV antibody model has been demonstrated in a number of studies of stress-induced immunosuppression (Table 7.1). Increased antibody titers have been associated with a wide range of stressors, including medical school exams (Glaser *et al.*, 1987, 1993), involvement in a poor quality marriage (Kiecolt-Glaser *et al.*, 1987a, 1988),

Table 7.1. *Previous research associating naturalistic stressors with EBV antibody titer*

Sample	Stressor	Association with EBV antibody	Reference
49 medical students	Major exams	↑	Glaser *et al.*, 1985
45 medical students			Glaser *et al.*, 1994
54 undergraduates	Self-reported anxiety or defensiveness	↑	Esterling *et al.*, 1993
38 married women	Poor quality marriage	↑	Kiecolt-Glaser *et al.*, 1987
32 married men			Kiecolt-Glaser *et al.*, 1988
76 women	Recent separation/divorce	↑	Kiecolt-Glaser *et al.*, 1987
64 men			Kiecolt-Glaser *et al.*, 1988
90 newlywed couples	Negative interactions	↑	Kiecolt-Glaser *et al.*, 1993
31 older married couples	during marital discussion		Kiecolt-Glaser *et al.*, 1997
29 West Point cadets	Basic training	←→	Glaser *et al.*, 1999
	Final exams	↑	
68 adults	Caring for family member with Alzheimer's disease	↑	Kiecolt-Glaser *et al.*, 1987b
65 gay men	Stress management intervention	↓	Esterling *et al.*, 1992
76 undergraduates	Disclosure of repressed trauma	↓	Lutgendorf *et al.*, 1994

and caring for a family member with Alzheimer's disease (Kiecolt-Glaser *et al.*, 1987b). Additionally, loneliness, defensiveness, and anxiety have all been positively associated with EBV antibody titers (Glaser *et al.*, 1985; Esterling *et al.*, 1993). Conversely, stress management interventions and disclosure of previously repressed trauma have been associated with decreases in EBV antibodies (Esterling *et al.*, 1992; Lutgendorf *et al.*, 1994). Furthermore, in comparison with other measures of immunity, meta-analysis has identified EBV antibodies as among the strongest and most consistent correlates of chronic stress (Herbert and Cohen, 1993). Although relatively few studies in psychoneuroimmunology make explicit links to disease outcomes (Keller *et al.*, 1994), relevant outcomes for the EBV antibody model include symptoms of infectious disease, neoplastic disease, and rates of wound healing.

Like memory responses to other previously encountered antigens, EBV-specific antibody titers can be expected to rise two to seven days after viral antigen re-exposure (Kuby, 1994). As such, the duration of time elapsing between stressor and EBV antibody response is on the order of days or weeks. EBV antibody levels are therefore not subject to short-term fluctuation, acute context effects, or diurnal variation, and a single sample can be used as an immunological measure of chronic stress. This is an advantage over other stress measures which require multiple samples, or are sensitive to the circumstances under which they are collected.

Any attempt to investigate the direct effects of psychosocial stress on immune function must pay careful attention to potential confounders such as nutritional status and current infection, both of which can independently modulate immunity (Kiecolt-Glaser and Glaser, 1988). The majority of research in PNI sidesteps this issue by recruiting only healthy individuals, but this may not always be the case for field-based research that includes larger, more representative samples from a wider range of populations. Recent research has also suggested that sleep quality may be a potential mediator of stress-immune relationships (Hall *et al.*, 1998).

Sample collection, transport, and storage

The requirements for collecting and processing blood-spot samples are relatively minimal (Figure 7.2). First, the participant's finger is cleaned with isopropyl alcohol, and then pricked with a sterile, disposable lancet that is commonly used by diabetics to monitor blood glucose

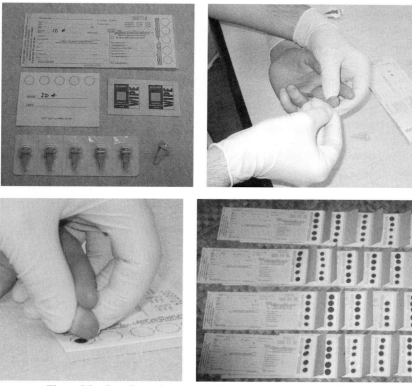

Figure 7.2.   Supplies and procedure for collecting finger prick blood spot samples.

(Microtainer, Franklin Lakes, NJ). The lancet is designed to deliver a controlled, uniform puncture that will stimulate sufficient capillary blood flow with minimal injury. The first drop of blood is wiped away, and up to five drops (~50 μL per drop) are spotted onto standardized filter paper (Whatman #903, Florham Park, NJ). The filter paper is certified to meet performance standards for sample absorption and lot-to-lot consistency set by the National Committee on Clinical Laboratory Standards, and by the Food and Drug Administration regulations for Class II Medical Devices. The samples are allowed to dry (four hours to overnight), at which point they can be stacked and stored in sealed plastic bags or plastic containers with desiccant.

This is a relatively painless, non-invasive, and convenient protocol for collecting and processing blood samples, particularly in comparison with venipuncture. Samples do not need to be centrifuged, separated, or immediately frozen, and samples can be stored and transported in airtight containers at ambient temperatures (although elevated temperatures should be avoided). In addition, finger-prick blood sampling can be performed by non-medically trained personnel, and samples can be collected virtually anywhere. Infants, children, and the elderly can provide blood without great difficulty, and repeat sampling becomes more feasible.

Requirements for shipping blood-spot samples are relatively minimal unless the samples are known to contain an infectious or etiologic agent. Samples from normal, healthy individuals are considered "diagnostic specimens," and according to current regulations, can be shipped without special packaging, labeling, or permitting. With respect to importing samples from overseas, the Centers for Disease Control and Prevention (CDC) will issue importation permits, although such permits may only be required under certain circumstances. Up-to-date shipping and importation guidelines are available from the CDC (http://www.cdc.gov/od/ohs/biosfty/biosfty.htm).

Prior work has shown that investigators can attain the same degree of precision and reliability with blood spots that they achieve with standard methods of sample collection such as venipuncture (Mei *et al.*, 2001). However, important sources of potential error should be acknowledged. First, proper placement of whole blood on the filter paper is essential. The uniform absorbing properties of the filter paper will be defeated if blood is blotted or smeared onto the paper, or if a drop of blood is placed on top of a previously collected drop. In addition, the volume of whole blood applied to filter paper as a blood spot can influence the volume of serum contained within a single disk punched out of that spot (Adam *et al.*, 2000). For this reason, an attempt should be made to ensure that all blood-spots used for analyses are of a comparable size. Variation in blood-spot size can be minimized by collecting samples on filter papers with pre-printed circles as guides (available from Schleicher & Schuell) to standardize the volume of whole blood collected from each individual. When filled to its border, each circle will contain approximately 50 μL of whole blood.

The filter paper matrix stabilizes most analytes in dried blood spots, but the rate of sample degradation will vary by analyte and should be evaluated prior to sample collection. To determine the stability of EBV antibodies in dried blood spots, a set of samples was exposed to one of three temperature conditions (4°C, room temperature, 37°C) for varying lengths of time up to eight weeks (McDade *et al.*, 2000a). Matched "baseline" samples served as the basis for comparison, and were frozen at

−23°C immediately after collection and drying. EBV antibodies were found to be stable in dried blood spots for at least eight weeks if stored at room temperature or at 4°C. However, samples began to deteriorate after one week of storage at 37°C. Blood spot samples collected for analysis of EBV antibodies should therefore be protected from excessive heat exposure. While it is always advisable to refrigerate or freeze samples when possible to minimize the chances of degradation, the stability of EBV antibodies in dried blood spots at room temperature provides flexibility in the collection of samples in field settings. The impact of repeated freezing and thawing on sample stability was also evaluated, with no evidence of deterioration in EBV antibody concentrations after six cycles of freeze/thaw.

Laboratory analysis

The dried blood-spot EBV antibody protocol is an adaptation of a commercially available enzyme-linked immunosorbent assay (ELISA) for measuring IgG antibodies against the EBV viral capsid antigen (VCA) complex (McDade *et al.*, 2000a). Beyond the materials provided with the assay kit, equipment requirements include access to a microplate spectrophotometer and a microplate washer, as well as equipment standard to most wet labs: deionized water, single channel and repeater pipettors, incubator, and facilities for biohazardous waste disposal.

The day before an assay is to be performed, blood-spot samples are removed from the freezer, and a small hole punch (available from office supply stores) is used to punch out one 3.2 mm disk of whole blood. The disk is transferred to a small glass test tube and incubated overnight in buffer, during which time the dried whole blood elutes from the disk and into solution. This reconstituted sample is then added in duplicate to microtiter wells. Antigen−antibody complexes form between EBV-VCA IgG antibodies present in the sample and synthetic peptide p18 bound to the surface of the wells. Horseradish peroxidase-labeled anti-human IgG in the presence of a chromogen substrate reacts with the antigen−antibody complex resulting in color development. The concentration of EBV-VCA IgG antibodies is directly related to the absorbance of the solution measured at 450 nm. Sample values are interpolated from a standard curve using a linear data-reduction protocol, and reported in ELISA units.

The accurate determination of previous exposure to EBV, and sub-sequent seroconversion, is critical since the model linking stress to suppressed cell-mediated immune function and increased EBV antibody level does not apply to seronegative individuals. Analyses must therefore focus exclusively on seropositive individuals. Previous comparison of matched plasma and blood-spot samples for seronegative and sero-positive individuals established a blood-spot seropositivity cut-off value of 20 ELISA units (McDade *et al.*, 2000a). Individuals with EBV antibody levels below this value are assumed to be seronegative for EBV, and must be excluded from analysis.

Previous analysis of EBV antibody concentration in a set of matched plasma and blood-spot samples indicates a high level of agreement across these two methods (Pearson R = 0.97, N = 40) (McDade *et al.*, 2000a). In addition, the blood-spot method has good precision and reliability, demonstrating within-assay and between-assay percent coefficients of variation of less than 10 percent. The assay for quantifying EBV antibodies in dried blood spots has been well validated, and represents a viable alternative to methods requiring serum or plasma.

Measuring inflammation: C-reactive protein (CRP) in dried blood spots

The CRP model

C-reactive protein is an inflammatory protein produced by liver hepatocytes in response to messenger cytokines, primarily IL-6 (Ballou and Kushner, 1992; Pepys, 1995). CRP is an important component of innate immunity, and has been used clinically for decades as an indicator of active infection. Recent development of high sensitivity CRP assays has led to the discovery that slight elevations of CRP − in the range of what was previously considered normal − are indicative of low-grade inflammatory processes that may be related to the pathophysiology of cardiovascular disease.

More than a dozen population-based studies have demonstrated that elevated CRP at baseline is a significant predictor of future cardiovas-cular disease, even after adjusting for other standard risk factors (Ridker *et al.*, 1998; Lagrand *et al.*, 1999; Libby *et al.*, 2002; Danesh *et al.*, 2004). While prior practice often employed 10 mg/L as a reference value for identifying active inflammatory conditions, these recent studies suggest that cardiovascular risk may increase significantly with baseline concentrations of CRP greater than 1−2 mg/L (Pearson *et al.*, 2003).

Research on the impact of psychosocial stress on inflammatory pathways is in its early stages, with evidence thus far indicating that this may be an important area for future work (Black, 2002). A number of studies have reported significant associations between symptoms of depression and markers of inflammatory activity, including IL-6, TNFα, and CRP (Dentino *et al.*, 1999; Danner *et al.*, 2003; Glaser *et al.*, 2003). Measures of burnout, self-reported distress, and post-traumatic stress disorder have also been linked with inflammation (Lutgendorf *et al.*, 1999; Maes *et al.*, 1999; Grossi *et al.*, 2003, McDade *et al.*, 2006), as has the chronic stress associated with caring for a relative with dementia (Kiecolt-Glaser *et al.*, 2003). Increased concentrations of CRP have also been reported in individuals of low socio-economic status, independent of a number of potentially confounding health behaviors (Owen *et al.*, 2003). Downstream health outcomes relevant to research on stress and inflammation include symptoms of cardiovascular and metabolic disease.

A recent review of indicators of inflammation conducted by the American Heart Association (AHA) and the Centers for Disease Control and Prevention (CDC) has recommended the limited measurement of CRP in clinical practice, and called for additional population-based research (Pearson *et al.*, 2003). They also propose cutpoints of low risk (< 1.0 mg/L), average risk (1.0−3.0 mg/L), and high risk (> 3.0 mg/L) that approximate the tertile distribution of serum/plasma CRP in a range of populations. Given the relative stability of CRP concentrations in individuals across time (Danesh *et al.*, 2004), a single measure can provide meaningful information on the level of chronic inflammation. Serum/plasma concentrations of CRP greater than 10 mg/L are assumed to represent an acute inflammatory response, likely to an infectious agent. The AHA/CDC guidelines therefore recommend discarding these values, and collecting another sample two weeks later that will provide a better indicator of chronic inflammation (Pearson *et al.*, 2003).

This raises a number of issues that should be considered prior to the application of blood-spot CRP to field-based research on stress and inflammation. First, repeat sampling is probably not feasible in population research given that it is highly unlikely that samples will be analyzed within two weeks of collection. Therefore, if the AHA/CDC guidelines are followed, a number of individuals with high CRP will have to be removed from the sample, possibly introducing bias. Second, the AHA/CDC cut-off values are based on concentrations of CRP measured in serum or plasma, and these concentrations will differ significantly in blood-spot samples. This can be addressed by using blood-spot CRP cut-off values based on population-specific tertiles, or by converting blood-spot results to plasma equivalents using previously established conversion factors (McDade *et al.*, 2004). Lastly, smoking, obesity, and intense physical activity are associated with increased CRP, and may confound the association between stress and inflammation.

Sample collection, transport, and storage

The requirements for collecting and processing blood-spot samples for CRP analysis are identical to those for EBV antibodies. The only exception is that CRP degrades more quickly, and therefore requires a higher degree of protection from heat exposure than samples used only for analysis of EBV antibodies. In an evaluation of sample integrity over a 14-day period, it was determined that CRP remains stable for at least 14 days at room temperature or at 4°C, but deteriorates significantly after three days at 37°C, or three days in an oscillating condition of 12 hours at 32°C and 12 hours at 22°C (simulating diurnal temperature variation in tropical environments) (McDade *et al.*, 2004). As with EBV antibodies, blood-spot CRP is robust to cycles of freeze/thaw, with no deterioration detected after five cycles.

Seven 3.2-mm disks can be punched out of a single full-size drop of whole blood (~50 μL) collected on filter paper. Since the CRP and EBV antibody assays each require only one disk, both assays can be performed from a single drop, with sample to spare. However, it is advisable to collect more than one drop in case samples need to be re-analyzed, and to allow for measurement of additional analytes in the future.

Laboratory analysis

Blood spot CRP is assayed using a high-sensitivity enzyme immunoassay protocol developed from commercially available reagents (McDade *et al.*, 2004). Equipment requirements include a microplate reader and washer, a hematology centrifuge, pH meter, and standard wet lab resources.

In comparison with the EBV antibody assay, the blood-spot CRP protocol requires a higher degree of technical proficiency in the lab, primarily due to the fact that blood-spot standards are made by adding washed erythrocytes to a serum-based reference preparation with known concentration of CRP. Washed erythrocytes are obtained by collecting several milliliters of whole blood (EDTA-anticoagulated) from a single individual, centrifuging at 1500 × g for 15 minutes, removing plasma and buffy coat, and rinsing the remaining erythrocytes in normal saline. This procedure is repeated three times, washed erythrocytes are added to the CRP reference preparation, mixed, and then spotted onto filter paper in 50 uL aliquots. Standards are then dried and stored at −30°C. Adding washed erythrocytes to the CRP reference preparation maximizes comparability between assay standards and the whole blood matrix of blood-spot samples. CRP standards can be prepared in batches in advance of sample analysis, and do not need to be made prior to each assay.

The day before an assay is to be performed, blood-spot standards and samples are removed from the freezer, and one 3.2-mm disk of whole blood is eluted in buffer overnight at 4°C. Also the day before (or further in advance in batches, if desired), a microtiter plate is coated overnight with anti-human CRP antibodies. The day of the assay, blood-spot eluate is pipetted in duplicate into microtiter wells. Through a series of incubation and wash steps, an antigen–antibody "sandwich" forms, with CRP from the sample located between peroxidase-conjugated anti-human CRP detection antibody and the capture antibody bound to the microtiter plate. Chromogen substrate reacts with this complex, resulting in color development proportional to the concentration of CRP. Absorbance is read at 490 nm, and CRP concentrations are calculated from the best-fit four-parameter logistic standard curve.

The performance characteristics of the blood-spot CRP assay have been previously evaluated, and indicate that the assay has good sensitivity, precision, and reliability (McDade *et al.*, 2004). The lower detection limit of the assay is 0.028 mg/L, well below the level needed to detect concentrations of CRP associated with cardiovascular disease risk. Assay performance was further evaluated by comparing CRP concentrations in 94 paired blood-spot and serum samples. Serum samples were analyzed with a widely used clinical assay system (IMMAGE[TM], Beckman Coulter, Inc.). The association between the blood-spot and serum methods is linear and the correlation is high (Pearson R = 0.96).

Measuring immune function in the field

The first field application of the blood-spot EBV antibody method was conducted in rural North Carolina, as part of the ongoing Great Smoky Mountains Study of adolescent psychopathology and service use (McDade *et al.*, 2000a). Finger-prick blood-spot samples were already being collected as part of the in-home interview, and a subsample of 256 9–13 year olds were selected to pilot the EBV antibody method. Of these individuals, 80.1% were seropositive for EBV. Fifty-one individuals did not show evidence of prior EBV exposure (19.9%), and therefore had to be excluded from subsequent analyses. This is a significant limitation of the EBV antibody method that may introduce a degree of bias, and that limits the generalizability of results to the population of EBV-positive individuals. Fortunately, the likelihood of seropositivity increases with age, thereby reducing the number of individuals that must be excluded.

Following previous research in psychoneuroimmunology, we hypothesized that negative life events would be associated with reduced cell-mediated immunity, as indicated by higher EBV antibody titers. Negative life events included the death of a close family member or friend, physical or sexual abuse, and exposure to a potentially traumatic experience (e.g. violence or accident). On average, participants had one negative life event in their lifetime ($SD = 1.2$, range $= 0–6$). Life events were not significantly associated with EBV antibodies in boys. However, for girls there was a strong positive association between negative life events and EBV antibodies (McDade *et al.*, 2000a). These analyses were the first to demonstrate an association between stress and EBV antibodies in a community-based setting.

Following these favorable results, the EBV antibody method was implemented in a non-Western field setting for the first time in the islands of Samoa. Samoa was selected due to its high rates of adolescent suicide – suggesting an exceptionally high level of adolescent distress – and due to its low rates of malnutrition and endemic disease, two factors that may obscure stress–immune function relationships. The objectives of the study were threefold: 1) evaluate the utility of the blood-spot EBV antibody method for measuring stress and immunity in a remote setting; 2) consider sources of stress for adolescents in Samoa and investigate their physiological impact; and 3) introduce a cross-cultural, population-level perspective to current research in psychoneuroimmunology. Analyses from this dataset have led to a series of papers investigating the social and cultural ecology of adolescent stress in Samoa (McDade *et al.*, 2000b; McDade, 2001, 2002; McDade and Worthman, 2004), but the discussion here is limited to life events, a dominant paradigm in current stress research.

It was hypothesized that negative life events would be associated with reduced immune function as previous research in Western populations has demonstrated, but that the significance of these events would be defined by the local cultural context (McDade, 2003). Participants were recruited from 14 villages across Samoa, and 295 individuals between the ages of 10 and 20 years provided data for the study, including a finger-prick blood-spot sample. Blood spots were collected in a centralized location in each village, dried overnight, stored in a refrigerator for up to two weeks, and then shipped via express mail to the USA. Of the 295 participants, only two had to be removed due to lack of prior EBV exposure, reflecting a 99% seropositivity rate for this population.

Measures of anthropometric status indicated that individuals in the study were well-nourished, and no significant relationships were found with EBV antibody level (McDade *et al.*, 2000b). To verify this finding in the subsample of adolescents used in the analysis of life events, body mass index and skinfold measures were evaluated as predictors of immune function. Nutritional status was not significantly associated with EBV antibody concentration or the number of life events, indicating that it was not likely to confound the association between stress and immunity.

A less sensitive version of the blood-spot CRP assay described above was used as a screening tool to identify individuals with a current or recent infection. CRP is a central component of the acute phase response, the body's first line of defense against infectious disease, and circulating concentrations increase by a factor of 100 to 1 000 during the 24 to 72 hours following an injury or infectious challenge (Fleck, 1989; Baumann and Gauldie, 1994). The half-life of circulating CRP is approximately 18 hours, and concentrations remain elevated during the course of infection for approximately one week following resolution (Gillespie *et al.*, 1991; Mortensen, 1994). Since CRP has been shown to increase in response to a wide range of pathogenic agents, it is a potentially useful marker of infectious burden. Of course, other methods may be applied to evaluate infection status, including physical exams and/or symptom diaries or recalls, although the CRP method has the advantage of being objective and less subject to recall bias.

Based on previous work in this population (McDade *et al.*, 2000b), a blood-spot CRP concentration greater than or equal to 5 mg/L was used as a cut-off to identify individuals with a current or recent infection. Participants with elevated CRP (N = 15) were removed from the sample prior to analysis to minimize the possibility of confounding. This is a conservative step that biases results toward the null, as individuals with current infection may be the same individuals who are suffering from the infectious consequences of stress-induced immunosuppression.

A summary life-events variable was constructed based on the following events occurring in the preceding year: death of a family member or close friend; serious illness in the family; hospitalization of a family member; and the number of *faalavelave* hosted by the family (0 = none, 1 = one, 2 = two or more). *Faalavelave* is a system of public, formalized gift exchange associated with community events such as weddings, funerals, and building dedications (Shore, 1982; O'Meara, 1990). Previous ethnographic work, as well as observations in the field based on a series of in-depth, semi-structured interviews, indicated that *faalavelave* can be a significant source of stress, particularly because it taxes a family's economic as well as social resources.

Life events were significantly related to EBV antibody titer in interaction with socioeconomic status (defined by parental occupation and remittances from relatives overseas). As expected, the association between life events and EBV antibody level was positive, such that adolescents with more life events had elevated antibody levels, indicating reduced cell-mediated immunity. This association was strongest for adolescents from families with low and middle levels of socioeconomic resources, whereas life events were not associated with suppressed immunity for adolescents with high resources. Figure 7.3 presents

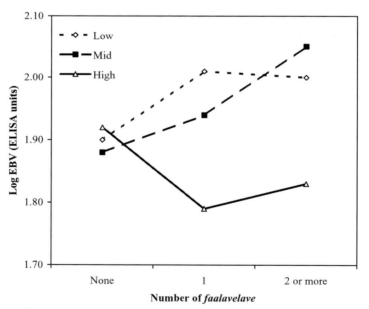

Figure 7.3. Interaction between number of *faalavelave* and socio-economic status in predicting EBV antibody titer in Samoan adolescents.

these results for *faalavelave* — the most frequently reported life event in the sample. For families with low economic resources, the costs of *faalavelave* may be particularly burdensome, forcing them to draw on the resources of extended kin networks in order to maintain their standing within the community. The economic and social debts that follow are likely to make these experiences more stressful for all members of the household.

These results are consistent with previous work in Western populations linking life events to suppressed immune function (Herbert and Cohen, 1993), and they demonstrate the feasibility and utility of measuring EBV antibodies and CRP in dried blood spots. These methods have since been applied to samples drawn from populations in Kenya, Bolivia, Russia, Kentucky, and Chicago.

Future directions

A constraint of blood-spot methods is the relatively small volume of the sample, which may limit the range of analytes that can be measured. For example, recent work suggests that patterns of cytokine expression may be important mediators of the associations among stress, immune function, and disease, but current assay protocols require volumes of blood (i.e., 50 or 100 µL of serum/plasma per cytokine) that are unattainable with blood spots. However, recent innovations in immunoassay technology allow the simultaneous quantification of multiple analytes in one sample, whereas previously only one analyte could be analyzed at a time. This advance is made possible by a multi-analyte flow analyzer, which uses different sets of polystyrene microspheres — pre-coated with a range of capture antibodies — to assay multiple analytes from a single sample.

Recent work has successfully used this method to measure up to 15 cytokines in less than a single drop of human plasma (Carson and Vignali, 1999; Cook *et al.*, 2001). The increased sensitivity, reduced cost, and low sample volume requirements afforded by this technology promise to expand the range of factors that can be measured in whole blood spots. Application of this method is in its early stages, but its feasibility with blood spots has already been demonstrated (Bellisario *et al.*, 2000). Much validation work remains, but this approach may facilitate the implementation of measures of cytokine activity into field-based research on stress and immunity.

Conclusion

Dried blood spots provide a "field-friendly" option for collecting blood samples that minimizes participant inconvenience and burden, and eases requirements associated with sample collection, transportation, and storage. Two measures of immune activity – EBV antibodies and CRP – can be reliably measured in blood spots, with additional measures possible in the future, subject to assay development and validation. These methods make possible population-level research on stress and immunity that can complement current laboratory and clinic-based approaches.

Advantages of the blood-spot EBV antibody and CRP methods include a minimally invasive sampling protocol that can be implemented in even the most remote field settings. By bringing these methods to research participants, the impact of stress on immune function can be evaluated in larger, more diverse, and more representative samples. An additional advantage of the EBV antibody method is that a single measure has been shown to be among the strongest immunological correlates of chronic stress. As such, it provides insight into the effects of psychosocial processes on a critical physiological system, and it can be used to complement other physiological stress indicators (e.g. blood pressure, cortisol).

A disadvantage of all blood-spot methods is that most research is conducted with serum or plasma samples, and comparisons of results across methods must be made with caution. However, the high correlation between blood-spot and serum/plasma methods for assessing EBV antibodies and CRP makes this less of a concern. Another disadvantage is the fact that EBV antibodies provide only a single measure of cell-mediated immune activity, and CRP provides only a single measure of inflammation. A more complete picture of the impact of stress on immune function will require additional measures. Lastly, a limitation of the EBV antibody method is that it relies on prior exposure to EBV, and high rates of seronegativity will limit statistical power and may introduce bias. However, due to the ubiquity of EBV, this is only likely to be a problem in samples of younger children from affluent settings.

Research in North Carolina and Samoa has demonstrated the feasibility of population-level research on stress and immune function, and future work should consider developing new field measures, including specific cytokines, thymic peptides, and antibody responses to vaccine challenge. With an expanding range of methodological options, field-based research on the social ecology of immune function will yield important insights into the physiological and health implications of stress for individuals in the USA and overseas.

Resource list

Sample collection

Lancets (#366357, Becton Dickinson, or equivalent)
Filter papers (#903, Whatman, Florham Park, NJ)

EBV antibody analysis

ELISA kit (P001606A, DiaSorin, Stillwater, MN)

CRP analysis

Calibrator (#X0923, Dako, Carpinteria, CA)
Coating antibody (#A0073, Dako)
Detection antibody (#P227, Dako)
Microtiter plate (#439454, NUNC Maxisorp)
Chromogen substrate (#S2045, Dako)

References

Adam, B. W., Alexander, J. R., Smith, S. J. *et al.* (2000). Recoveries of phenylalanine from two sets of dried-blood-spot reference materials: prediction from hematocrits, spot volume, and paper matrix. *Clinical Chemistry*, **46**, 126–8.

Affleck, G., Urrows, S., Tenne, H. *et al.* (1997). A dual pathway model of daily stressors effects on rheumatoid arthritis. *Annals of Behavioral Medicine*, **19**, 161–70.

Ballou, S. P. and Kushner, I. (1992). C-reactive protein and the acute phase response. *Advances in Internal Medicine*, **37**, 313–36.

Baumann, H. J. and Gauldie, J. (1994). The acute phase response. *Immunology Today*, **15**(2), 74–80.

Bellisario, R., Colinas, R. J. and Pass, K. A. (2000). Simultaneous measurement of thyroxine and thyrotropin from newborn dried blood-spot specimens using a multiplexed fluorescent microsphere immunoassay. *Clinical Chemistry*, **46**(9), 1422–4.

Biondi, M. (2001). Effects of stress on immune functions: an overview. In *Psychoneuroimmunology*, 3rd edn, eds. R. Ader, D. L. Felten and N. Cohen. San Diego CA: Academic Press, pp. 189–226.

Biondi, M., Costantini, A. and Parisi, A. M. (1996). Can loss and grief activate latent neoplasia? *Psychotherapy and Psychosomatics*, **65**, 102–5.

Black, P. H. (2002). Stress and the inflammatory response: a review of neurogenic inflammation. *Brain, Behavior, and Immunity*, **16**, 622–53.

Boyce, W. T., Chesney, M., Alkon, A. *et al.* (1995). Psychobiologic reactivity to stress and childhood respiratory illnesses: results of two prospective studies. *Psychosomatic Medicine*, **57**, 411–22.

Boyce, W. T., Chesterman, E. A., Martin, N. *et al.* (1993). Immunologic changes occurring at kindergarten entry predict respiratory illnesses after the Loma Prieta earthquake. *Journal of Developmental and Behavioral Pediatrics*, **14**, 296–303.

Boyce, W. T., Jensen, E. W., Cassel, J. C. *et al.* (1977). Influence of life events and family routines on childhood respiratory tract illness. *Pediatrics*, **60**(4), 609–15.

Carson, R. T. and Vignali, D. A. A. (1999). Simultaneous quantitation of 15 cytokines using a multiplexed flow cytometric assay. *Journal of Immunological Methods*, **227**, 41–52.

Cohen, S. (1995). Psychological stress and susceptibility to upper respiratory infections. *American Journal of Respiratory Critical Care Medicine*, **152**(Supplement), 53–8.

Cohen, S., Frank, E., Doyle, W. J. *et al.* (1998). Types of stressors that increase susceptibility to the common cold in healthy adults. *Health Psychology*, **17**(3), 214–23.

Cook, E. B., Stahl, J. L., Lowe, L. *et al.* (2001). Simultaneous measurement of six cytokines in a single sample of human tears using microparticle-based flow cytometry: allergics vs. non-allergics. *Journal of Immunological Methods*, **254**, 109–18.

Cook, J. D., Flowers, C. H. and Skikne, B. S. (1998). An assessment of dried blood-spot technology for identifying iron deficiency. *Blood*, **92**, 1807–13.

Danesh, J., Wheeler, J. G., Hirschfield, G. M. *et al.* (2004). C-reactive protein and other circulating markers of inflammation in the prediction of coronary heart disease. *New England Journal of Medicine*, **350**, 1387–97.

Danner, M., Kasl, S. V., Abramson, J. L. and Vaccarino, V. (2003). Association between depression and elevated C-reactive protein. *Psychosomatic Medicine*, **65**, 347–56.

Dentino, A. N., Pieper, C. F., Rao, K. M. K. *et al.* (1999). Association of interleukin-6 and other biologic variables with depression in older people living in the community. *Journal of the American Geriatrics Society*, **47**, 6–11.

Erhardt, J. G., Craft, N. E., Heinrich, F. and Bielaski, H. K. (2002). Rapid and simple measurement of retinol in dried whole blood spots. *Journal of Nutrition*, **132**(2), 318–21.

Esterling, B. A., Antoni, M. H., Kumar, M. and Schneiderman, N. (1993). Defensiveness, trait anxiety, and Epstein-Barr viral capsid antigen antibody titers in healthy college students. *Health Psychology*, **12**(2), 132–9.

Esterling, B. A., Antoni, M. H., Schneiderman, N. *et al.* (1992). Psychosocial modulation of antibody to Epstein-Barr viral capsid antigen and human

Herpesvirus Type-6 in HIV-1-infected and at-risk gay men. *Psychosomatic Medicine*, **54**, 354–71.

Fleck, A. (1989). Clinical and nutritional aspects of changes in acute-phase proteins during inflammation. *Proceedings of the Nutrition Society*, **48**, 347–54.

Gillespie, S. H., Dow, C., Raynes, J. G. *et al.* (1991). Measurement of acute phase proteins for assessing severity of *Plasmodium falciparum* malaria. *Journal of Clinical Pathology*, **44**, 228–31.

Glaser, R., MacCallum, R. C., Laskowski, B. F. *et al.* (2001). Evidence for a shift in the Th1 to Th2 cytokine response associated with chronic stress and aging. *Journal of Gerontology A. Biological Sciences and Medical Sciences*, **56**, M477–82.

Glaser, R., Kiecolt-Glaser, J. K., Bonneau, R. G. *et al.* (1992). Stress-induced modulation of the immune response to recombinant hepatitis B vaccine. *Psychosomatic Medicine*, **54**, 22–9.

Glaser, R., Kiecolt-Glaser, J. K., Speicher, C. E. and Holliday, J. E. (1985). Stress, loneliness, and changes in herpesvirus latency. *Journal of Behavioral Medicine*, **8**(3), 249–60.

Glaser, R., Pearl, D. K., Kiecolt-Glaser, J. K. and Malarkey, W. B. (1994). Plasma cortisol levels and reactivation of latent Epstein-Barr virus in response to examination stress. *Psychoneuroendocrinology*, **19**(8), 765–72.

Glaser, R., Pearson, G. R., Bonneau, R. H. *et al.* (1993). Stress and the memory T-cell response to the Epstein-Barr virus in healthy medical students. *Health Psychology*, **12**(6), 435–42.

Glaser, R., Pearson, G. R., Jones, J. F. *et al.* (1991). Stress-related activation of Epstein-Barr virus. *Brain, Behavior, and Immunity*, **5**, 219–32.

Glaser, R., Rice, J., Sheridan, J. *et al.* (1987). Stress-related immune suppression: health implications. *Brain, Behavior and Immunity*, **1**, 7–20.

Glaser, R., Friedman, S. B., Smyth, F. *et al.* (1999). The differential impact of training stress and final examination stress on herpesvirus latency at the United States Military Academy at West Point. *Brain, Behavior and Immunity*, **13**, 240–51.

Glaser, R., Robles, T. F., Sheridan, J., Malarkey, W. B. and Kiecolt-Glaser, J. K. (2003). Mild depressive symptoms are associated with amplified and prolonged inflammatory responses after influenza virus vaccination in older adults. *Archives of General Psychiatry*, **60**, 1009–14.

Grossi, G., Perski, A., Evangard, B., Blomkvist, V. and Orth-Gomer, K. (2003). Physiological correlates of burnout among women. *Journal of Psychosomatic Research*, **55**, 309–16.

Hall, M., Baum, A., Buysse, D. J. *et al.* (1998). Sleep as a mediator of the stress-immune relationship. *Psychosomatic Medicine*, **60**, 48–51.

Henle, W. and Henle, G. (1982). Epstein-Barr virus and infectious mononucleosis. In *Human Herpesvirus Infections: Clinical Aspects*, ed. R. Glaser and T. Gottleib-Stematsky. New York: Marcel Dekker, 151–62.

Herbert, T. B. and Cohen, S. (1993). Stress and immunity in humans: a meta-analytic review. *Psychosomatic Medicine*, **55**, 364–79.

Ironson, G., Wynings, C., Schneiderman, N. *et al.* (1997). Posttraumatic stress symptoms, intrusive thoughts, loss, and immune function after Hurricane Andrew. *Psychosomatic Medicine*, **59**, 128–41.

Irwin, M., Daniels, M., Smith, T. L., Bloom, E. and Weiner, H. (1987). Impaired natural killer cell activity during bereavement. *Brain, Behavior, and Immunity*, **1**, 98–104.

Jemmott, J. G. and Magloire, K. (1988). Academic stress, social support, and secretory immunoglobulin A. *Journal of Personality and Social Psychology*, **55**(5), 802–10.

Keller, S. E., Shiflett, S. C., Schleifer, S. J. and Bartlett, J. A. (1994). Stress, immunity, and health. In *Handbook of Human Stress and Immunity*, ed. R. Glaser and J. Kiecolt-Glaser. San Diego, CA: Academic Press, 217–44.

Kiecolt-Glaser, J. K., Fisher, L. D., Ogrocki, P. *et al.* (1987a). Marital quality, marital disruption, and immune function. *Psychosomatic Medicine*, **49**(1), 13–34.

Kiecolt-Glaser, J. K. and Glaser, R. (1988). Methodological issues in behavioral immunology research with humans. *Brain, Behavior, and Immunity*, **2**, 67–78.

(1995). Psychoneuroimmunology and health consequences: Data and shared mechanisms. *Psychosomatic Medicine*, **57**, 269–74.

(1997). Measurement of immune response. In *Measuring Stress: A Guide for Health and Social Scientists*, ed. S. Cohen, R. C. Kessler and L. U. Gordon. New York: Oxford University Press, 213–29.

Kiecolt-Glaser, J. K., Glaser, R., Gravenstein, S., Malarkey, W. B. and Sheridan, J. (1996). Chronic stress alters the immune response to influenza virus vaccine in older adults. *Proceedings of the National Academy of Sciences*, **93**, 3043–7.

Kiecolt-Glaser, J. K., Glaser, R., Shuttleworth, E. C. *et al.* (1987b). Chronic stress and immunity in family caregivers of Alzheimer's disease victims. *Psychosomatic Medicine*, **49**, 523–35.

Kiecolt-Glaser, J. K., Kennedy, S., Malkoff, S. *et al.* (1988). Marital discord and immunity in males. *Psychosomatic Medicine*, **50**, 213–29.

Kiecolt-Glaser, J. K., Malarkey, W. B., Cacioppo, J. T. and Glaser, R. (1994). Stressful personal relationships: immune and endocrine function. In *Handbook of Human Stress and Immunity*, ed. R. Glaser and J. K. Kiecolt-Glaser. San Diego, CA: Academic Press, 321–39.

Kiecolt-Glaser, J. K., Marucha, P. T., Malarkey, W. B., Mercado, A. M. and Glaser, R. (1995). Slowing of wound healing by psychological stress. *Lancet*, **346**, 1194–6.

Kiecolt-Glaser, J. K., Glaser, R., Gravenstein, S., Malarkey, W. B. and Sheridian, J. (1996). Chronic stress alters the immune response to influenza virus vaccine in older adults. *Proceedings of the National Academy of Sciences*, **93**, 3043–7.

Kiecolt-Glaser, J. K., Glaser, R., Cacioppo, J. T. and Malarkey, W. B. (1998). Marital Stress: immunologic, neuroendocrine, and autonomic correlates. *Annals of the New York Academy of Sciences*, **840**, 656–63.

Kiecolt-Glaser, J. K., Preacher, K. J., MacCallum, R. C. *et al.* (2003). Chronic stress and age-related increases in the proinflammatory cytokine IL-6. *Proceedings of the National Academy of Sciences*, **100**, 9090–5.

Kuby, J. (1994). *Immunology*, 2nd edn. New York: W.H. Freeman.

Lagrand, W. K., Visser, C. A., Hermens, W. T. *et al.* (1999). C-reactive protein as a cardiovascular risk factor: more than a epiphenomenon? *Circulation*, **100**, 96–102.

Libby, P., Ridker, P. M. and Maseri, A. (2002). Inflammation and athero-sclerosis. *Circulation*, **105**, 1135–43.

Lutgendorf, S. K., Antoni, M. H., Kumar, M. and Schneiderman, N. (1994). Changes in cognitive coping strategies predict EBV-antibody titre change following a stressor disclosure induction. *Journal of Psychosomatic Medicine*, **38**(1), 63–78.

Lutgendorf, S. K., Garand, L., Buckwalter, K. C. *et al.* (1999). Life stress, mood disturbance, and elevated interleukin-6 in healthy older women. *Journal of Gerontology: Medical Sciences*, **54A**, M434–9.

Maes, M., Lin, A., Delmeire, L. *et al.* (1999). Elevated serum interleukin-6 (IL-6) and IL-6 receptor concentrations in posttraumatic stress disorder following accidental man-made traumatic events. *Biological Psychiatry*, **45**, 833–9.

Marshall, G. D., Agarwal, S. K., Lloyd, C. (1998). Cytokine dysregulation associated with exam stress in healthy medical students. *Brain, Behavior, and Immunity*, **12**, 297–307.

Marucha, P. T., Kiecolt-Glaser, J. K. and Favagchi, M. (1998). Mucosal wound healing is impaired by examination stress. *Psychosomatic Medicine*, **60**, 362–5.

McDade, T. W. (2001). Lifestyle incongruity, social integration, and immune function in Samoan adolescents. *Social Science and Medicine*, **53**, 103–14.

  (2002). Status incongruity in Samoan youth: a biocultural analysis of culture change, stress, and immune function. *Medical Anthropology Quarterly*, **16**, 123–50.

  (2003). Life event stress and immune function in Samoan adolescents: toward a cross-cultural psychoneuroimmunology. In *Social and Cultural Lives of Immune Systems*, ed. J. Wilce. New York: Routledge, 170–88.

McDade, T. W., Burhop, J. and Dohnal, J. (2004). High sensitivity enzyme immunoassay for C-reactive protein in dried blood spots. *Clinical Chemistry*, **50**, 652–4.

McDade, T. W. and Shell-Duncan, B. (2002). Whole blood collected on filter paper provides a minimally-invasive method for assessing transferrin receptor level. *Journal of Nutrition*, **132**, 3760–3.

McDade, T. W., Stallings, J. F., Angold, A. *et al.* (2000a). Epstein-Barr virus antibodies in whole blood spots: A minimally-invasive method for assessing an aspect of cell-mediated immunity. *Psychosomatic Medicine*, **62**, 560–67.

McDade, T. W., Stallings, J. F. and Worthman, C. M. (2000b). Culture change and stress in Western Samoan youth: methodological issues in the cross-cultural study of stress and immune function. *American Journal of Human Biology*, **12**, 792–802.

McDade, T. W. and Worthman, C. M. (2004). Socialization ambiguity in Samoan adolescents: a new model for research in human development and stress in the context of culture change. *Journal of Research in Adolescence*, **14**, 49–72.

McDade, T. W., Hawkley, L. C. and Cacioppo, J. T. (2006). Psychosocial and behavioral predictors of inflamation in middle-age and older adults: the Chicago health, aging, and social relations study. *Psychosomatic Medicine*, **68**, 376–81.

Mei, J. V., Alexander, J. R., Adam, B. W. and Hannon, W. H. (2001). Use of filter paper for the collection and analysis of human whole blood specimens. *Journal of Nutrition*, **131**, 1631–6S.

Meyer, R. J. and Haggerty, R. J. (1962). Streptococcal infections in families: Factors altering individual susceptibility. *Pediatrics*, **29**, 539–49.

Mortensen, R. F. (1994). Macrophages and acute-phase proteins. *Immunology Series*, **60**, 143–58.

Moynihan, J. A., Kruszeska, B., Brenner, G. J. and Cohen, N. (1998). Neural, endocrine, and immune system interactions. Relevance for health and disease. *Advances in Experimental Medicine and Biology*, **438**, 541–9.

O'Meara, J. T. (1990). *Samoan Planters: Tradition and Economic Development in Polynesia*. Fort Worth, TX: Holt, Rinehart and Winston.

Owen, N., Poulton, T., Hay, F. C., Mohamed-Ali, V. and Steptoe, A. (2003). Socioeconomic status, C-reactive protein, immune factors, and responses to acute mental stress. *Brain, Behavior, and Immunity*, **17**, 286–95.

Pearson, T. A., Mensah, G. A., Alexander, R. W. *et al.* (2003). Markers of inflammation and cardiovascular disease: application to clinical and public health practice. *Circulation*, **107**, 499–511.

Pepys, M. B. (1995). The acute phase response and C-reactive protein. In *Oxford Textbook of Medicine Vol. 2*, 3rd edn, ed. D. J. Weatherall, J. G. G. Ledingham and D. A. Warrell. Oxford: Oxford University Press, 1527–33.

Petrie, K. J., Booth, R. J., Pennebaker, J. W. and Davison, K. P. (1995). Disclosure of trauma and immune response to a Hepatitis B vaccination program. *Journal of Consulting and Clinical Psychology*, **63**(5), 787–92.

Pike, J. L., Smith, T. L., Hauger, R. L. *et al.* (1997). Chronic life stress alters sympathetic, neuroendocrine, and immune responsitivity to an acute psychological stressor in humans. *Psychosomatic Medicine*, **59**(4), 447–57.

Ridker, P. M., Buring, J. E., Shih, J., Matias, M. and Hennekens, C. H. (1998). Prospective study of C-reactive protein and the risk of future cardiovascular events among apparently healthy women. *Circulation*, **98**, 731–3.

Ross, R. (1999). Atherosclerosis: An inflammatory disease. *New England Journal of Medicine*, **340**, 115–26.

Schleifer, S. J., Scott, B., Stein, M. and Keller, S. E. (1986). Behavioral and developmental aspects of immunity. *Journal of the American Academy of Child Psychiatry*, **26**(6), 751–63.

Segerstrom, S. C., Kemeny, M. E. and Laudenslager, M. L. (2001). Individual difference factors in psychoneuroimmunology. In *Psychoneuroimmunology*, 3rd edn, eds. R. Ader, D. L. Felten and N. Cohen. San Diego, CA: Academic Press, 87–109.

Shore, B. (1982). *Sala'ilua: A Samoan Mystery*. New York: Columbia University Press.

Solomon, G. F., Segerstrom, S. C., Grohr, P., Kemeny, M. and Fahey, J. (1997). Shaking up immunity: Psychological and immunologic changes after a natural disaster. *Psychosomatic Medicine*, **59**, 114–27.

Tracy, R. P. (1998). Inflammation in cardiovascular disease. *Circulation*, **97**, 2000–2.

Uchino, B. N., Cacioppo, J. T. and Kiecolt-Glaser, J. K. (1996). The relationship between social support and physiological processes: a review with emphasis on underlying mechanisms and implications for health. *Psychological Bulletin*, **119**(3), 488–531.

Worthman, C. M. and Stallings, J. F. (1994). Measurement of gonadotrophins in dried blood spots. *Clinical Chemistry*, **40**(3), 448–53.

(1997). Hormone measures in finger-prick blood spot samples: New field methods for reproductive endocrinology. *American Journal of Physical Anthropology*, **104**(1), 1–22.

Zautra, A. J., Hoffman, J., Potter, P. *et al.* (1997). Examination of changes in interpersonal stress as a factor in disease exacerbations among women with rheumatoid arthritis. *Annals of Behavioral Medicine*, **19**, 279–86.

Part III
Practical issues in studying stress

8 Measuring stress in special populations

SHARON R. WILLIAMS

In the introduction to this volume, Ice and James refer to "practical considerations" as an important component when selecting a measure of stress. This chapter will discuss some practical considerations in detail in the consideration of stress measurement in reproductive age women, children, older populations, and in non-clinical settings. Selection of culturally relevant questionnaires and interview techniques in field settings are discussed in detail elsewhere in this volume, so this chapter will focus on the collection of physiological markers of stress in non-clinical settings. Current research has suggested new areas of interest and new biological markers of relevance in these special populations. Some practical considerations discussed in this chapter include selection of relevant and appropriate biological markers of stress in special populations and logistical considerations associated with collecting biological markers in non-clinical settings. Many of these factors can impact compliance rates, the number of individuals who will agree to participate and successfully complete the study as well as the acceptability of biological samples for laboratory analysis.

Stress and reproducing women

Stress and reproductive function

According to the demographer Bongaarts (1983), there are seven "proximate determinants" of fertility that mediate all other variables (e.g. social, economic, and cultural) that control the length of birth intervals. Stress has been suggested to impact reproduction through all of Bongaart's proximate determinants of fertility (Bongaarts, 1983). Consequences of Sympathetic Adrenal Medullary System (SAMS) and/ or Hypothalamic Pituitary Adrenal (HPA) axis activation have been identified as interfering with exposure factors (i.e. age at menarche, coital frequency), waiting time to conception (reproductive cycles, ovulation), pregnancy loss, and gestation length (parturition). Hormones associated with human stress responses, including corticotrophin releasing hormone (CRH), adrenocorticotropic hormone (ACTH) and cortisol affect the HPA axis which controls reproductive function (Calogero *et al.*, 1998; Veldhuis *et al.*, 1998). Once a neurological stress response has been initiated, the production of stress hormones is stimulated in the hypothalamus, pituitary and adrenal cortex. An increase in any of these hormones has been found to negatively impact the GnRH pulse generator located in the hypothalamus and may impact reproduction (Veldhuis *et al.*, 1998). Restraint stress, immune stress and stress associated with transportation have all been directly associated with reduction in pulsatile leutenizing hormone (LH) and follicle stimulating hormone (FSH) secretion in many animal models (Breen *et al.* 2005). CRH has also been found to suppress ovarian steroid production in vitro (Calogero *et al.*, 1996; Ghizzoni *et al.*, 1997; Erden *et al.*, 1998)

The impact of psychosocial stress on human reproductive cycles and ovulation is not clear; in fact, contradictory results have been reported. Some research suggests that since "emotional upsets" and chronic anxiety do not contribute to long-term amenorrhea, psychosocial stress is not a significant factor in determining fertility (Schacter and Shoham, 1994; Bringer *et al.*, 1997). Other studies have found no significant impact of stress-related factors on fertility and fecundity. For example, physiological indicators of stress were not significantly different between menstruating and non-menstruating long distance runners (Loucks and Horvath, 1985) and stress levels were not related to probability of conception in a sample of working Danish women (Hjollund *et al.*, 1999). In the USA and Europe, psychosocial and emotional stress levels are associated with amenorrhea and infertility in some women. According to Berga *et al.* (2003), cortisol is elevated in women suffering from amenorrhea when compared to normal menstruating women and stress-reduction techniques and psychological intervention increase chances of conception in these women. A similar association between stress levels and likelihood of conception was reported in a large sample of Danish women (Hjollund *et al.*, 1998, 1999). Women with long menstrual cycles who reported higher levels of distress, based on a standardized questionnaire, were less likely to conceive than their less distressed counterparts. Several other studies have also concluded that in well-nourished Western populations, psychosocial distress may be a risk factor for reduced fertility (Veldhuis *et al.*, 1998; Sanders and Bruce, 1999).

Stress and pregnancy

During pregnancy, stress associated with increased levels of corticotrophin releasing hormone (CRH), has recently been linked to multiple negative outcomes. CRH plays important roles in the neuroendocrine, behavioral, autonomic, and immune responses to stress (Herman *et al.*, 1996) and is secreted from the hypothalamus to act on the pituitary to increase secretion of ACTH and other pro-opiomelanocortin products (Kasckow *et al.*, 2001). Receptors for CRH are found not only in the pituitary, but also in the ovaries, placenta, and testes (Perrin and Vale, 1999). CRH during pregnancy, produced both by the placenta and by the maternal hypothalamus, increases steadily during pregnancy and peaks just prior to the onset of labor (Amiel-Tison *et al.*, 2004a,b). In cases of infection or hypoxia, the placenta has been identified as initiating the rise in CRH and potentially the onset of parturition (Amiel-Tison *et al.*, 2004a). In addition to its role in parturition, CRH triggers the release of fetal cortisol to stimulate organ development (Challis *et al.*, 2001; Mulder *et al.*, 2002). While short-term increases in stress and the CRH driven cascade can result in acceleration of fetal development (Amiel-Tison *et al.*, 2004a), long-term increases in CRH and associated hormones will result in fetal distress, potentially associated with pregnancy loss, intrauterine growth retardation, and short gestation (Amiel-Tison *et al.*, 2004a; Mulder *et al.*, 2002).

These clinically negative outcomes, when considered from an evolutionary perspective, have been suggested as adaptive. From the maternal perspective, the costs and benefits of continuing a pregnancy under stressful circumstances as well as the biological consequences of continued gestation and potential consequences to future reproductive events must be considered (Stearns, 1992; Pike, 2005). The costs and benefits to the fetus must also be considered: whether the chance of survival is better in utero, depending on stage of development and in utero conditions (Amiel-Tison *et al.*, 2004a).

The immediate benefit of survival through pregnancy with the alterations of early delivery or adjusted fetal growth can have long-term consequences. Low birth-weight babies (<2500 g) and preterm infants (<37 weeks of gestation) have a higher risk of poor perinatal and infant outcomes (Johnston *et al.*, 2001; Kramer, 2003), the increased risk of morbidity and mortality may continue across the life span (Kramer, 1987; Barker *et al.*, 1992; Amiel-Tison *et al.*, 2004b; Seckl and Meaney, 2004). Increased stress during pregnancy has also been associated with more difficult infant behavior (deWeerth *et al.*, 2003; Austin *et al.*, 2005).

Further, there may be direct influence on programming of the fetal stress response (Meaney, 2001) and even remodeling of the human brain (Welberg and Seckl, 2001; de Kloet and Oitzl, 2003).

Measuring stress in women, infants and children

The biological links between external stressors and reproduction in women and the impacts created on offspring by these biological responses are interesting and important, but often difficult to examine outside of a clinical setting. However, variation in these systems cannot be fully understood from clinical data alone. Large-scale data collections as well as collection from non-Western populations are important for understanding the adaptive value of these stress responses in differing environments. Practical considerations are important to consider, both in terms of acceptability to a specific population and suitability of the biological specimen for storage, transport, and analysis, depending on specific circumstances.

An important practical consideration in measurement selection when dealing with women of reproductive age is minimization of participant burden. The necessity of repeated measures across the menstrual cycle, across cycles, and during pregnancy makes the use of blood particularly burdensome in the study of the biological consequences of stress in women. Stress levels and stress responses can change over the course of the menstrual cycle, across menstrual cycles and throughout a pregnancy. Further complicating the collection of biological measures of stress and fecundity in women is the high amount of variation in the menstrual cycle length and timing of ovulation both within and between women (Wood, 1994).

Saliva and urine are less burdensome to collect in regular intervals, however some markers of interest are either not available in saliva or urine or reliable assays have not yet been developed. Both ACTH and CRH are measured in serum or plasma and are extremely difficult to collect under non-clinical circumstances. Alternate methods for analysis of these proteins have not been developed. The half-life of ACTH is only 12–17 minutes; blood samples must be collected, centrifuged and frozen immediately after collection and must remain frozen during transporation. CRH is much more stable when collected by venipuncture; blood-spot assays are not currently available.

Measuring stress in children and adolescents is a difficult and challenging process, not only because of the unique and ever-changing stressors associated with these stages of life, but also due to the additional methodological challenges of this population. In very young subjects, blood is often difficult to obtain and emotionally difficult for both parent and child. Compliance with urine collection methods can be equally difficult. Diapers can aid in the collection of urine, but it can be difficult to extract the urine from the diaper. Remembering to collect multiple urine samples can also be difficult for young children.

Saliva is the least invasive and causes the least amount of discomfort for young populations. However, standard techniques for measuring cortisol from saliva, such as the use of cotton rolls or spitting through tubes, are not useful in very young subjects. Alternative methods have been developed to deal with these subjects. Sterile cotton, similar to that used in other collection methods, can be manually swabbed in the mouths of infants, or alternately, thick cotton twine. After sufficient moistening, these tools can be centrifuged to separate the saliva for assay. In addition, several types of suction devices can be used to manually remove saliva from the mouth. Small, sterile, disposable plastic serological pipettes (for example, VWR cat# 101093–966) or needleless syringes (for example, VWR cat# BD301503) are useful tools for manual saliva collection. Both are relatively inexpensive (currently less than $0.50/piece) and readily available from most lab suppliers.

Stress in older populations

There is ample evidence that suggests that the response of both the Sympathetic Adrenal Medullary System (SAMS) and the Hypothalamic Pituitary Adrenal (HPA) axis change with aging. Decreasing sensitivity of these systems to stress and delayed return to baseline levels following elevation have been reported in older populations. However, the human aging process is far from homogenous; in fact, human variation in phenotype increases dramatically with age (McEwen, 1999; MacLullich *et al.*, 2005).

Stress, cognitive function and the brain

The source of dysregulation of the human stress response with age is most likely in the human brain, specifically the hippocampus. Evidence from both animal models and human observations suggests that the human brain is plastic. Remodeling of the brain takes place in response to physical and psychological stress. Over time this adaptive response may result in the dysregulation of the SAMS and HPA as well as other systems such as memory and cognitive function. As with the rest of the aging process, changes in memory and cognitive function with age vary significantly from individual to individual. A growing body of literature suggests that stress and the dysregulation of the stress systems can be related to cognitive decline and loss of memory (Brosschot *et al.*, 2005; McEwen, 2005). However, the relationship is very complex with both risk factors and protective factors that have not yet been identified.

The hippocampus is the center for declarative/spatial learning and memory in the human brain and is neurologically and anatomically connected to the hypothalamus. The hippocampus also responds to stress; specifically, levels of glucocorticoids. Sapolsky *et al.* (1986) suggest that glucocorticoid exposure over time can lead to damage such as neuronal death, structural change and decline in function which would promote progressive elevation of adrenal steroids and eventual dysregu-lation of the HPA axis. Further work on both animal models and humans suggests that the response of the hippocampus to stress, and any resulting damage, depends not just on glucocorticoids, but also excitatory amino acids, neurotropins and calcium channels (McEwen, 1999).

The conversion of plasticity to permanent damage and permanent decline in cognitive function is not yet understood. Important questions still remain as to whether chronic stress, acute stress associated with negative life events, prenatal stress exposure result and/or interactions between these types of stress result in permanent damage. Gender differences in the incidence of cognitive impairment and Alzheimer's disease are also present, but not well understood. Further, differences in the measurement of cognitive function confound the study of the relationship between stress and cognitive function. Cognition includes memory, attention, and visual/motor performance. Tools for measuring cognition are vast and easily available. Some basic tests, suitable for field studies and their specific measures are provided in Table 8.1.

Stress, aging and gender

Further confounding factors, contributing to the variability in the aging phenotype, are apparent gender differences in the HPA axis response to stress with aging. Although most research suggests that there are no differences in the HPA stress response in young adults (Kudielka and Kirschbaum, 2005), there is strong evidence in the literature suggesting

Table 8.1. *Measures of cognitive function*

	Measures	Reference
Mini Mental Status Examination (MMSE)	concentration, language, memory	Folstein *et al.*, 1975
Short Portable Mental Status Questionnaire (SPMSQ)	knowledge of general and personal information	Pfeiffer, 1975
Wechsler Memory Scale	memory	Wechsler, 1987
Rey Auditory–Verbal Learning Test	immediate/delayed recall	Lezak, 1995
Benton Visual Retention Test (BVRT)	non-verbal reasoning	Sivan, 1992
Raven's Standard Progressive Matrices	non-verbal reasoning	Raven *et al.*, 1977
Cognitive Abilities Screening Instrument (CASI)	verbal learning and memory, clock drawing	Borson *et al.*, 2000

that there are significant differences between male and female responses to stress in aging populations. This evidence, however, is inconsistent. For example, Seeman *et al.* (1995) reported women were more likely to have high cortisol responses to a driving challenge but Traustadottir *et al.* (2003) and Kudielka *et al.* (2004) reported higher cortisol responses to a battery of mental tests in men. In a recent meta-analysis of previously published studies, Otte *et al.* (2005) reported finding a significant increase in cortisol responses in older women over men.

Estrogen, and its interaction with the HPA axis, has been suggested as a protective factor against stress-induced damage in the brain and has also implicated stress as a factor in the development of breast cancer. Both endogenous and exogenous estrogen (hormone replacement) have been associated with decreased risk of the development of cognitive impairment and Alzheimer's disease in animal studies, clinical trials and prospective cohort studies (Savaskan, 2001). Rasgon *et al.* (2005) found that increased reproductive period (resulting in an increase in estrogen exposure) significantly decreased risk of cognitive impairment in a study of Swedish twins. Other research has indicated that hormone replacement provided protection from cognitive decline in women (Kawas *et al.*, 1997; Zec and Trivedi, 2002; Pinkerton and Henderson, 2005).

The relationship between estrogen, stress, and breast cancer has been examined in several studies, with conflicting results. Estrogen is thought to play both a direct role, through tumor initiating effects, and an indirect role, through promotion of prolactin and growth factor secretion, in the development of breast cancer (Nandi *et al.*, 1995). Although estrogen from either endogenous or exogenous sources has been linked to breast cancer (Clemons and Goss, 2001), only endogenous estrogen exposure is affected by the HPA axis. Suppression of estrogen production by activation of the HPA over the course of the life cycle, through multiple acute stressors or chronic stress, can decrease overall cumulative exposure of breast tissue to estrogen.

Studies of recent general perception suggest that most individuals and healthcare professionals believe that stress increases the risk of breast cancer (Baghurst *et al.*, 1992; Steptoe and Wardle, 1994). However, most research into the association between psychosocial stress and risk of breast cancer has found no relationship between self-reported daily stress (Butow *et al.*, 2000; Duijts *et al.*, 2003; Kroenke *et al.*, 2004; Schernhammer *et al.*, 2004) or acute stress (Lambe *et al.*, 2004) and increased risk of breast cancer. Conversely, other research has suggested that higher levels of estrogen in postmenopausal women (Toniolo *et al.*, 1995; Cauley *et al.*, 1999) were associated with increased risk of breast cancer; higher levels of self-reported daily stress were associated with lower levels of estrogen (Kroenke *et al.*, 2004); and higher rates of self-reported daily stress were actually associated with a decreased risk of breast cancer development (Nielsen *et al.*, 2005). To further complicate this relationship, recent research suggests that women who have a close relative who has developed breast cancer tend to exhibit higher levels of reported emotional distress and increased physiological response to stress in both cortisol and epinephrine secretion (Gold *et al.*, 2003; James *et al.*, 2004; Dettenborn *et al.*, 2005).

Measuring stress in older populations

Level of both cognitive and physical function are important to consider when selecting measures of stress in older populations outside of clinical settings. Mobility limitations in older populations as well as cognitive function may limit the ability for reliable self collection of salivary samples for cortisol analysis. Reduced cognitive function can make compliance difficult, even when samples are not self-collected. Cues, such as watches, timers and palm pilots, successfully used in other age groups, may be difficult to use in this population due to vision loss, reduced cognitive function, or simply unease with new technology.

Traditional venipuncture blood draws may also be more challenging in older populations because stable veins are difficult to find after frequent blood draws, some chemotherapy, and certain diseases. Capillary blood collection would be a viable alternative to venipuncture under these circumstances. For saliva collection, the use of salivettes, cotton rolls or alternative methods used in children, may be preferred over straws and tubes for collection of saliva in those with limited hand function. Some medications can decrease saliva production and make the use of stimulants appealing. Estrogen can also be readily and reliably assayed in saliva, as well as in plasma, serum, urine and dried blood spots. As in children, urine can be difficult to collect from older adult participants who are incontinent. A good understanding of the characteristics and needs of the older population of interest is vital in success measurements of older adults.

Measuring stress in non-clinical settings

Stress research in non-clinical settings can include anything from the study of stress in community or home-based studies in the USA to very remote populations in rural parts of the developing world. Many of the restraints and practical concerns are the same. Biological markers have a benefit in non-traditional settings in that they are not culturally specific. This is not to say they are culturally irrelevant, but they are a part of the stress response in all human populations. The method of collection, however, may be very culturally sensitive. Biological markers of stress can be collected from physical measures such as blood pressure as well as almost every body fluid. Ability or desire to comply with the methods associated with collection will vary with populations. Further, different populations may react differently to requests for venous vs. capillary blood or saliva vs. urine.

Other practical considerations include transportability or access to necessary equipment and sample processing, storage or transportation. Venous blood must be collected by trained individuals. Centrifuging samples shortly after collection and then freezing the plasma or serum until assay is the current standard practice. Dried blood spots have an obvious advantage over venous blood collection in that they can be easily collected by minimally trained researchers and do not need to be centrifuged or frozen for transport. However, assays currently available and validated for use with dried whole blood are limited and, as discussed in Chapter 7, comparison of values assayed from whole blood spots and those from plasma or serum must be done with care. Development of new assays or the adaptation of currently available kits for use with dried blood spots are needed before this method of collection can replace venipuncture in the study of the biological stress response.

   Timing of saliva collection can also be difficult in non-clinical settings. Saliva collected with the aid of cotton also must be centrifuged but saliva collected with a straw does not. All saliva, however, needs to be frozen if assays are not performed immediately, to prevent bacterial growth that will interfere with some assays. In some cases, preservatives such as sodium azide may be used in the absence of freezing, but these preservatives may also interfere with some assays. In the case of cortisol, if individuals cannot collect and store their own samples, careful consideration must be given to the method and timing of sample collection in order to compensate for the diurnal pattern of cortisol secretion. Individuals may be called to a central area at a given time, but this requires deviation from normal daily activities and may result in increased or decreased stress levels. Travel from household to household for collection may only be feasible if the area is densely populated or reliable transportation is available. The same considerations for prevention of bacterial growth also apply to urine samples.

   Distance and amount of time required for transportation from location of collection to location of assay also merits consideration. Shipment within the USA of blood (including blood spots, plasma and serum), saliva and urine require samples to be labeled and shipped as "diagnostic specimens." These samples must be packed with enough absorbable material to collect any sample leakage and packed in a sealed, leak-proof bag or container. This container must be packed inside an additional sealed container clearly marked as "diagnostic specimens" on the outside of the package. Diagnostic specimens can be shipped through commercial carriers or through standard US mail when correctly packaged (contact the USDOT or commercial shipper for details). Commercially available coolers and reusable cold packs can be used to keep samples cold or frozen for short or long term shipments. Dry ice may also be used for shipping within the USA but additional packing and labeling regulations apply to the shipment of dry ice (for details contact the USDOT). Similar requirements apply to international shipment of diagnostic specimens with the addition of export regulations of the country of origin. In addition, depending on the sample source and location of origin, additional permits for import of samples into the USA may be required from the Centers for Disease Control (contact the CDC for details — http://www.cdc.gov/od/ohs/biosfty/0753.pdf). Failure to comply with CDC requirements (requirements vary, but may include informing participants and local governments of the export of samples) may result in a customs delay or sample destruction on return to the USA.

Conclusions

The special populations discussed in this chapter are those populations for which very little information is available on the impact of stress and health. While collection of valid information, both biological and social, is more challenging in these populations, clinically derived data provides only a small part of the picture by focusing primarily on white, Western, adult populations, and not reflecting real-world, everyday situations. Consideration of the methodology for collection of the measure of interest and the burden associated with collection for the participant must be done from a cultural context. Decisions about methods and markers are often practical ones, limited by circumstances, money, and time rather than ideal protocols. Although physiological markers of stress were the focus of this chapter, the physiological response must be connected to the external stressor through the individual in order for the human stress response to be fully understood.

References

Amiel-Tison, C., Cabrol, D., Denver, R. *et al.* (2004a). Fetal adaptation to stress part I. Acceleration of fetal maturation and earlier birth triggered by placental insufficiency in humans. *Early Human Development*, **78**, 15–27.

(2004b). Fetal adaptation to stress part II. Evolutionary aspects: Stress-induced hippocampal damage; long-term effects on behavior; consequences on adult health. *Early Human Development*, **78**, 81–94.

Austin, M., Hadzi-Pavlovic, D., Leader, L., Saint, K. and Parker, G. (2005). Maternal trait anxiety, depression and life event stress in pregnancy: Relationships with infant temperament. *Early Human Development*, **81**, 183–90.

Baghurst, K. I., Baghurst, P. A. and Record, S. J. (1992). Public perceptions of the role of dietary and other environmental factors in cancer causation and prevention. *Journal of Epidemiology and Community Health*, **46**, 120–6.

Barker, D. J. (1992). The fetal origins of adult hypertension. *Journal of Hypertension*, **10**, S39–44.

Berga, S. L., Marcus, M. D., Loucks, T. L. *et al.* (2003). Recovery of ovarian activity in women with functional hypothalamic amenorrhea who were treated with cognitive behavior therapy. *Fertility and Sterility*, **80**, 976–81.

Bongaarts, J. (1983). The proximate determinants of natural marital fertility. In *Determinants of Fertility in Developing Countries*, ed. R. A. Bulatao and R. D. Lee. New York: Academic Press, pp. 108–38.

Borson, S., Scanlan, J., Brush, M., Bitaliano, P. and Dokmak, A. (2000). The mini-cog: a cognitive 'vital signs' measure for dementia screening in multi-lingual elderly. *International Journal of Geriatric Psychiatry*, **12**, 1021–7.

Breen, K. M., Billings, H. J., Wagenmaker, E. R., Wessinger, E. W. and Karsch, F. J. (2005). Endocrine basis for disruptive effects of cortisol on preovulatory events. *Endocrinology*, **146**, 2107–15.

Bringer, J., Lefebvre, P., Boulet, F., Clouet, S. and Renard, E. (1997). Deficiency of energy balance and ovulatory disorders. *Human Reproduction*, **12**, 97–109.

Brosschot, J. S., Pieper, S. and Thayer, J. F. (2005). Expanding stress theory: prolonged activation and preservative cognition. *Psychoneuroendocrinology*, **30**, 1043–9.

Butow, P. N., Hiller, J. E., Price, M. A. *et al.* (2000). Epidemiological evidence for a relationship between life events, coping style, and personality factors in the development of breast cancer. *Journal of Psychosomatic Research*, **49**, 169–81.

Calogero, A. E., Burrello, N., Negri-Cesi, P. *et al.* (1996). Effects of corticotrophin-releasing hormone on ovarian estrogen production in vitro. *Endocrinology*, **137**, 4161–6.

Calogero, A. E., Bagdy, G. and D'Agata, R. (1998). Mechanisms of stress on reproduction–evidence for a complex intra-hypothalamic circuit. *Annals of the New York Academy of Sciences*, **851**, 364–70.

Cauley, J. A., Lucas, F. L., Kuller, L. H. *et al.* (1999). Elevated serum estradiol and testosterone concentrations are associated with a high risk for breast cancer. *Annals of Internal Medicine*, **130**, 270–7.

Challis, J. R., Sloboda, D., Matthews, S. G. *et al.* (2001). The fetal placental hypothalamic–pituitary–adrenal (HPA) axis, parturition and post natal health. *Molecular and Cellular Endocrinology*, **185**, 135–44.

Clemons, M. and Goss, P. (2001). Estrogen and the risk of breast cancer. *New England Journal of Medicine*, **344**, 276–85.

de Kloet, E. R. and Oitzl, M. S. (2003). Who cares for a stressed brain? The mother, the kid, or both? *Neurobiology of Aging*, **24**, 61–5.

de Weerth, C., van Hees, Y. and Buitelaar, J. K. (2003). Prenatal maternal cortisol levels and infant behavior during the first 4 months. *Early Human Development*, **74**, 139–51.

Dettenborn, L., James, G. D., van Berge-Landry, H. *et al.* (2005). Heightened cortisol response to daily stress in working women at familial risk for breast cancer. *Biological Psychology*, **69**, 167–79.

Duijts, S. F. A., Zeegers, M. P. A. and Borne, B. V. D. (2003). The association between stressful life events and breast cancer risk: a meta-analysis. *International Journal of Cancer*, **107**, 1023–9.

Erden, H. F., Zwain, I. H., Asakura, H. and Yen, S. S. C. (1998). Corticotrophin-releasing factor inhibits luteinizing hormone-stimulated P450c17 gene expression and androgen production by isolated thecal cells of human ovarian follicles. *Journal of Clinical Endocrinology and Metabolism*, **83**, 448–52.

Folstein, M. F., Folstein, S. E. and McHugh, P. R. (1975). "Mini mental state." A practical method for grading the cognitive state of patients for the clinician. *Journal of Psychiatric Research*, **12**, 189–98.

Ghizzoni, L., Mastorakos, G., Vottero, A. *et al.* (1997). Corticotrophin-releasing hormone (CRH) inhibits steroid biosynthesis by cultured human granulosa-lutein in a CRH and interleukin-1 receptor mediated fashion. *Endocrinology*, **138**, 4806–11.

Gold, S. M., Zakowski, S. G., Baldimarsdottir, H. B. and Bovjerg, D. H. (2003). Stronger endocrine response after brief psychological stress in women at familial risk of breast cancer. *Psychoneuroendocrinology*, **28**, 584–93.

Herman, A. A., Barendes, H. W., Yu, K. F. *et al.* (1996). Evaluation of the effectiveness of a community-based enriched model prenatal intervention project in the District of Columbia. *Health Services Research*, **31**, 609–21.

Hjollund, N. H. I., Jensen, T. K., Bonde, J. P. E. *et al.* (1998). Job strain and time to pregnancy. *Scandinavian Journal of Work, Environment, and Health*, **24**, 344–50.

(1999) Distress and reduced fertility: a follow-up study of first-pregnancy planners. *Fertility and Sterility*, **72**, 47–53.

James, G. D., van Berge-Landry, H., Valdimarsdottir, H. B., Montgomery, G. H. and Bovbjerg, D. H. (2004). Urinary catecholamine levels in daily life are elevated in women at familial risk of breast cancer. *Psychoneuroendocrinology*, **29**, 831–8.

Johnston, R. J., Williams, M., Hogue, C. and Mattison, D. (2001). Overview: New perspectives on the stubborn challenge of preterm birth. *Paediatric and Perinatal Epidemiology*, **15**, 3–6.

Kasckow, J. W., Baker, D. and Geracioti, T. D. J. (2001). Corticotrophin-releasing hormone in depression and post-traumatic stress disorder. *Peptides*, **22**, 845–51.

Kawas, C., Resnick, S., Morrison, A. *et al.* (1997). A prospective study of estrogen replacement therapy and the risk of developing Alzheimer's disease: the Baltimore longitudinal study of aging. *Neurology*, **48**, 1517–21.

Kramer, M. S. (1987). Determinants of low birth weight: Methodological assessment and meta-analysis. *Bulletin of the World Health Organization*, **65**, 663–737.

(2003). The epidemiology of adverse pregnancy outcomes: an overview. *Journal of Nutrition*, **133**, 1592–6S.

Kroenke, C. H., Hankinson, S. E., Schernhammer, E. S. *et al.* (2004). Caregiving stress, endogenous sex steroid hormone levels, and breast cancer incidence. *American Journal of Epidemiology*, **159**, 1019–27.

Kudielka, B. M., Buske-Kirschbaum, A., Hellhammer, D. H. and Kirschbaum, C. (2004). HPA axis responses to laboratory psychosocial stress in healthy elderly adults, younger adults, and children: impact of age and gender. *Biological Psychology*, **69**, 113–32.

Kudielka, B. M. and Kirschbaum, C. (2005). Sex differences in HPA axis responses to stress: a review. *Biological Psychology*, **69**, 113–32.

Lambe, M., Cerrato, R., Askling, J. and Hsieh, C. (2004). Maternal breast cancer risk after the death of a child. *International Journal of Cancer*, **110**, 763–6.

Lezak, M. D. (1995). *Neuropsychological Assessment*, 3rd edn. New York: Oxford University Press.

Loucks, A. B. and Horvath, S. M. (1985). Exercise-induced stress responses and amenorrheic and eumenorrheic runners. *Journal of Clinical Endocrinology and Metabolism*, **59**, 1109–20.

MacLullich, A. M. J., Deary, I. J., Starr, J. M. *et al.* (2005). Plasma cortisol levels, brain volumes and cognition in healthy elderly men. *Psychoneuroendocrinology*, **30**, 505–15.

McEwen, B. S. (1999). Stress and hippocampal plasticity. *Annual Review of Neuroscience*, **22**, 105–22.

    (2005). Stress and the aging hippocampus. *Frontiers in Neuroendocrinology*, **20**, 49–70.

Meaney, M. J. (2001). Maternal care, gene expression, and the transmission of individual differences in stress reactivity across generations. *Annual Review of Neuroscience*, **24**, 1161–92.

Mulder, E. J. H., Robles de Medina, P. G., Huizink, A. C. (2002). Prenatal maternal stress: effects on pregnancy and the (unborn) child. *Early Human Development*, **70**, 3–14.

Nandi, S., Guzman, R. C. and Yang, J. (1995). Hormones and mammary carcinogenesis in mice, rats, and humans: a unifying hypothesis. *Proceedings of the National Academy of Sciences*, **92**, 3650–7.

Nielsen, N. R., Zhang, Z., Kristensen, T. S. *et al.* (2005). Self reported stress and risk of breast cancer: prospective cohort study. *British Medical Journal*, **331**, 548–53.

Otte, C., Hart, S., Neylan, T. C. *et al.* (2005). A meta-analysis of cortisol response to challenge in human aging: Importance of gender. *Psychoneuroendocrinology*, **30**, 80–91.

Perrin, M. H. and Vale, W. W. (1999). Corticotrophin releasing factor receptors and their ligand family. *Annals of the New York Academy of Sciences*, **885**, 312–28.

Pfeiffer, E. (1975). A short portable mental status questionnaire for the assessment of organic brain deficit in elderly patients. *Journal of the American Geriatrics Society*, **23**, 433–41.

Pike, I. L. (2005). Maternal stress and fetal responses: evolutionary perspectives on preterm delivery. *American Journal of Human Biology*, **17**, 55–65.

Pinkerton, J. V. and Henderson, V. W. (2005). Estrogen and cognition, with a focus on Alzheimer's disease. *Seminars in Reproductive Medicine*, **23**, 172–9.

Raven, J. C., Court, J. H. and Raven, J. (1977). *Manual for Raven's Progressive Matrices and Vocabulary Scales*. London: H. K. Lewis Company.

Rasgon, N. L., Magnusson, C., Johansson, A. L. V. *et al.* (2005). Endogenous and exogenous hormone exposure and risk of cognitive impairment in Swedish twins: a preliminary study. *Psychoneuroendocrinology*, **30**, 567.

Sanders, K. A. and Bruce, N. W. (1999). Psychosocial stress and the menstrual cycle. *Journal of Biosocial Science*, **31**, 393–402.

Sapolsky, R. M., Krey, L. and McEwen, B. S. (1986). The neuroendocrinology of stress and aging: the glucocorticoid cascade hypothesis. *Endocrinology Review*, **7**, 284–301.

Savaskan, E., Olivieri, G., Meier, F., Ravid, R. and Muller-Spahn, F. (2001). Hippocampal estrogen beta-receptor immunoreactivity is increased in Alzheimer's disease. *Brain Research*, **908**, 113–119.

Schacter, M. and Shoham, Z. (1994). Amenorrhea during the reproductive years – is it safe? *Fertility and Sterility*, **62**, 1–16.

Schernhammer, E. S., Hankinson, S. E., Rosner, B. *et al.* (2004). Job stress and breast cancer risk. *American Journal of Epidemiology*, **160**, 1079–86.

Seckl, J. R. and Meaney, M. J. (2004). Glucocorticoid programming. *Annals of the New York Academy of Science*, **1032**, 63–84.

Seeman, T. E., Singer, B. and Charpentier, P. (1995). Gender differences in patterns of HPA axis response to challenge: MacArthur studies of successful aging. *Psychoneuroendocrinology*, **20**, 711–25.

Sivan, A. B. (1992). *Benton Visual Retention Test*, 5th edn. New York: Psychological Corporation.

Steptoe, A. and Wardle, J. (1994). What the experts think: a European survey of expert opinion about the influence of lifestyle on health. *European Journal of Epidemiology*, **10**, 195–203.

Stearns, S. (1992). *The Evolution of Life Histories*. New York: Oxford University Press.

Toniolo, P. G., Levitz, M. and Zeleniuch-Jacquotte, A. (1995). A prospective study of the endogenous estrogens and breast cancer in postmenopausal women. *Journal of the National Cancer Institute*, **87**, 190–7.

Traustadottir, T., Bosch, P. R. and Matt, K. S. (2003). The HPA axis response to stress in women: effects of aging and fitness. *Psychoneuroendocrinology*, **30**, 392–402.

Veldhuis, J. D., Yoshida, K. and Iranmanesh, A. (1998). The effects of mental and metabolic stress on the female reproductive system and female reproductive hormones. In *Handbook of Stress Medicine*, ed. J. R. Hubbar and E. A. Workman. Boca Raton, FL: CRC Press, pp. 115–27.

Wechsler, D. (1987). *Wechsler Memory Scale – Revised, WMS-R*. New York: Psychological Corporation.

Welberg, L. A. M. and Seckl, J. R. (2001). Prenatal stress, glucocorticoids and the programming of the brain. *Journal of Neuroendocrinology*, **13**, 113–28.

Wood, J. (1994). *Human Reproductive Ecology*. New York: New York Academy of Science.

Zec, R. F. and Trivedi, R. A. (2002). The effects of ERT on neuropsychological function in postmenopausal women with and without dementia: a critical and theoretical review. *Neuropsychology Review*, **12**, 65–109.

9 *Study design and data analysis*

GARY D. JAMES AND GILLIAN H. ICE

Stress implies that a physiological or psychological change has occurred within a subject or that, because of a broader ecological stressor, measurable biological differences between subjects or groups of subjects exist. A variety of study designs can be employed to detect the effects of stress on biology and behavior in the field, although these variants fall under two general categories: 1) natural experiments with an a priori ecological framework and 2) model building or testing where there is no ecological framework per se, but rather a framework of expected relationships based on the results of prior field and/or laboratory research. Depending upon whether a field study of stress is designed to evaluate individual change or group differences, single or multiple measurements per subject can be evaluated.

The purpose of this chapter is to discuss, from a practical standpoint, research design and analytic approaches and techniques that are useful for field studies of stress. The intent is to provide researchers with information on how to develop a meaningful stress study and, hopefully, some insight into how best to evaluate stress-related data. It is not the intent of this chapter to detail specific statistical procedures; however, we will, when necessary, make reference to them.

Research design and the constraints of data collection

In conducting any study of stress response in the field (outside the laboratory), it is best to proceed with a plan in mind, one that will guide not only what data will be collected, but also one that is tied to the analytic technique (statistics) that will be used to test the study hypotheses. While there is often a tendency to collect as much data as possible while in the field, one should plan on collecting only those data that directly pertain to the hypotheses to avoid unnecessary effort and burden to the participants. In formulating the plan, the practical aspects of data collection must be considered. There are currently limitations in the technology and techniques available for stress studies in the field (such as with ambulatory blood pressure monitoring, saliva collection, urine collection and blood collection) that will determine how often and under what circumstances data collection can occur (see Chapters 2–7).

Sometimes, the best available research approach may actually have biases. The biases, however, can be used for advantage, if the researcher has given appropriate thought as to how they can be manipulated. Such biases usually emerge from practical constraints associated with the data collection, such as the non-randomness of the study sample, limitation of possible study groups (such as only being able to sample a single gender), unbalanced data collection (different numbers of measurements per study participant) or smallness of the sample size (number of participants). From a statistical perspective, biases can introduce type I or type II errors into the analysis, where a type I error would lead to rejecting a null hypothesis that should not be rejected, and a type II error would lead to failing to reject a null hypothesis that should be rejected (Table 9.1). Manipulating biases for advantage means that they are addressed in the study design in such a way that their effects tend to strengthen rather than compromise the results. Often this means that the hypothesis becomes less precise, such that the research may only determine whether an effect exists, rather than evaluating the magnitude of the effect.

An example of a study where bias was used to enhance the findings is that of James *et al.* (1986), which addressed the question of whether the experience of moods increased blood pressure in real life. In this study, participants wore an ambulatory blood pressure monitor that was preset to measure blood pressures every 15 minutes during waking hours. The subject also filled out a diary in which they reported the experience of anger, happiness, anxiety and sadness during each of the blood-pressure measurements. An underlying assumption of the study was that if the moods were to increase a subject's blood pressure during the day, the blood pressures taken when moods were reported should be higher than those taken when no moods were reported.

Table 9.1. *Type I and type II errors*

Decision	State of reality	
	H0 is true	H0 is false
Retain H0	Correct decision	Type II error (β)
Reject H0	Type I error (α)	Correct decision

Another way of stating this assumption is that if a distribution was developed for all the blood pressures taken on a subject, those occurring during a reported mood should be on the high end of the distribution.

In the study, there was a constraint related to the method of data collection. Specifically, study subjects reported moods only when and if they experienced them through the course of their normal daily activities. Consequently, each individual in the study could report a different mix and frequency of moods and mood intensities during their blood-pressure monitoring. In the study, some subjects were highly emotive on the day of monitoring, to the point where every blood pressure was taken during a reported mood state, while others were not moody at all, to the point where only one or two of their daily pressures were taken during a mood state. Thus, there was unbalanced data, in that there was a differential sampling of mood-associated blood pressures across subjects. From a statistical perspective, those subjects who were highly emotive (who had all their pressures taken during a mood) could dominate the analysis in that they contribute more measurements in the across-subject sample of emotion-associated blood pressures. For subjects whose every blood pressure was associated with a mood (sometimes the same mood), there was no differentiation between mood-associated and non-mood-related blood pressures. The entire distribution of pressures was mood associated.

This unbalanced data was likely to introduce a type II error in the analysis, in that the subjects with all or most of their blood pressure distribution associated with moods could dilute the average mood effects across subjects, potentially to a degree where the average mood-related pressures were equal to the individual's average daily pressure (the expected value if there was no mood effects). If this were the case, there would be a greater probability that the null hypothesis (mood-associated and non-mood-related pressures are equal) could not be rejected. However, with the data biased toward finding no effect, if significant mood effects were found, it would be clear that they would likely be larger than the calculated effects. Thus, even with the bias caused by differential mood experience, the hypothesis that moods increase blood pressure could still be confirmed if mood effects were found. In the study, different moods were found to increase blood pressure to differing degrees, and it was suggested that the estimated effects were probably smaller than the true effects. Subsequent studies, which have improved on mood assessment and used more sophisticated statistical procedures (as they have become available with statistical packages such as SAS and

SPSS since the study was conducted in 1986), have confirmed the findings of this seminal study, specifically that all mood experiences increase blood pressure to some extent, but that some moods (such as happiness) have less of an effect on increasing blood pressure than others (particularly anger) during real life (Schwartz *et al.*, 1994; Jacob *et al.*, 1999).

Design approaches in field studies

Aside from data-collection constraints, examining differences in the variation of biological stress responses outside the laboratory is further limited by the fact that it is not really possible to use experimental designs where control (baseline) situations can be introduced or where the conditions the subjects experience can be manipulated. Because there is interest in evaluating how real-life factors initiate stress responses, designs need to focus on the characteristics of the natural setting, or the study subjects themselves.

Natural experiments

In conducting "natural experiments" the researcher should look for naturally occurring environmental contrasts that can influence biological stress responses (Garruto *et al.*, 1999). Natural experiments allow the researcher to directly test differences or changes in the stress response associated with the naturally occurring stressor contrast. An assumption in these experiments is that the population groups in each of the settings would have similar distributions of measurements were it not for the stressor differences that define the contrasting settings. Some natural experiments provide a framework for comparing differences between genetically similar groups who live in environments that vary with regard to their stressfulness. The study of stress and modernization among sub-populations of Samoans (James *et al.*, 1985; McGarvey and Schendel, 1986; Pearson *et al.*, 1993; McDade *et al.*, 2000) was a good example of this between sub-population approach, where there were genetically similar populations living under conditions that differed with regard to exposure to westernized lifestyles. By examining and comparing bio-logical stress responses (blood pressure, catecholamines and later immune responses) both within and across these sub-groups it was possible to determine whether there was a difference in overall stressfulness of the

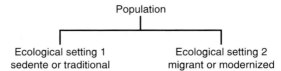

Figure 9.1. Natural experimental design highlighting between-group comparisons among sub-populations. There is a single measurement per subject, and the assumption behind this design is that the average measurement of each sub-population would be the same if it weren't for the ecological differences in the environments (sedente vs. migrant, or traditional vs. modernized) in which the sub-populations live.

lifestyles and to identify some of the characteristics that led to the varying responses (James *et al.*, 1985; McGarvey and Schendel, 1986; Pearson *et al.*, 1993; McDade *et al.*, 2000). Figure 9.1 illustrates the general form of this type of design.

One note of caution when using a natural experimental model (or any study design), is to be aware of the potential for confounding and a priori assumptions of what conditions are stressful. Confounding refers to the case when the association between two variables is due to a third variable which is associated with the two variables of interest. For example, in the case of stress in two genetically similar populations with different exposures to modernization, a researcher may find that blood pressure is higher in the modernized population, leading to the assumption that modernization is stressful. Blood pressure may be higher due to differences in diet or obesity levels across the two sub-populations, rather than differences in exposure to stressors. In this case, careful attention to measurement and control of potential confounding factors is critical. In addition, assumptions of what is stressful may be incorrect. In laboratory studies, participants are exposed to a stressor and those that react physiologically are considered reactors and those that do not are labeled non-reactors. However, as cognitive psychologists have pointed out, a stressor needs to be appraised as stressful to initiate a physiological reaction. The same is true for field studies. For example, in two separate studies of Samoans undergoing modernization (McGarvey and Baker, 1979; James *et al.*, 1985), the groups who were intermediate in their levels of acculturation had higher stress levels (as measured by blood pressure and catecholamines, respectively) than the most modernized group. Similarly, Brown (1981, 1982) found that Filipino-Americans in Honolulu who had intermediate levels of Americanization had the highest catecholamine levels. A simple comparison of modernized vs. traditional groups would have missed this fact.

12. week in
SNF → week
at DK home.

 Other natural experiments can occur when a study participant moves through different environments during the course of a day. By following the individual across these environments and measuring them in each environment, it is possible to evaluate how the environmental change affects biological stress responses and whether or not the individual is successfully adapting to the challenges presented by the changing settings (James, 1991). Examining biological measures using this type of natural experiment could also be viewed as a method of measuring the physiological process of allostasis, where the extent of the variability of physiological traits and their adaptive responses are examined (Sterling, 2004). This type of natural experimental approach is often used in studies that employ ambulatory blood pressure and/or urinary stress hormone levels (see, for example, Brown *et al.*, 2003; Ice *et al.*, 2003; Dettenborn *et al.*, 2005), and is discussed in some detail in Chapter 6. This type of design has, at its roots, laboratory experimental approaches that are often used in psychophysiological studies (see Chapter 6). Specifically, a given daily situation (such as sleep) is considered to be the "control" condition and biological responses to other daily situations, such as being at work or home are considered the experimental conditions. By examining the differences in biological responses between the situations, the effects of the stress associated with the situations can be deduced. Figure 9.2 illustrates this natural design approach.

Model testing

Model testing starts with a paradigm that is based on the results of prior studies from the laboratory or earlier field studies. A model is constructed to predict the biological or behavioral stress measurement (such as blood pressure, cortisol, reported anxiety, etc.). The independent variables in the model are parameters that have shown some kind of relationship with

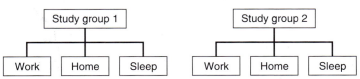

Figure 9.2.   Natural experimental design highlighting within-subject comparisons across microenvironments. There are multiple measurements per participant (one in each microenvironment) and the assumption behind this design is that the changes across the microenvironments would be the same if it weren't for the differences in the characteristics of the groups (physiological, genetic and/or cultural).

the stress response in prior studies. In testing hypothesized models of stress responses, the researcher measures, as best as can be done, the attributes that correspond to the independent variables in the model. Then, an analysis is done to determine how well the data collected fit the hypothesized model. In this approach if the data don't fit the model, the model may be adjusted to better fit the data. When that occurs, in essence, the a priori hypotheses about the relationships between the independent variables and the stress response measures are abandoned. A problem often confronted in model testing is that factors evaluated in prior studies are often highly correlated, meaning that they lack independence from one another; thus it may be difficult to determine which of the factors actually contributes to the variation in stress response when the correlated factors are both forced into the model.

Testing models based on experimental approaches is a way of extending and confirming the results of laboratory experiments under conditions where experimental manipulation is not possible. The previously noted studies of Schwartz *et al.* (1994) and Jacob *et al.* (1999) are excellent examples of studies that employed model testing to evaluate biological stress responses in the field. These studies examined the effects of moods on ambulatory blood pressures, and began with a framework that was developed from earlier laboratory and natural experimental studies (Schwartz *et al.*, 1981; James *et al.*, 1986).

Single vs. multiple measurements per participant

Natural experiments

Depending upon whether stress is being evaluated by contrasting populations under different settings or examined intra-individually by assessing biological or behavioral changes, there will be either one or multiple biological or behavioral measurements taken per subject. Natural experimental studies designed to examine between-subject variation and population distributions usually employ single measurements per subject. In these types of studies, there is an inference that contrasting sub-populations would show a similar average response were it not for the difference in a specific ecologically based stressor (McGarvey and Baker, 1979; McDade *et al.*, 2000). In evaluating ecologically based stressors, sub-populations may be chosen because they have migrated to a different cultural ecological setting (such as in modernization/migration studies) or because they differ with regard to specific characteristics *in situ*, such as their degree of isolation from other populations (Garruto *et al.*, 1999).

In natural experiments where stress-related sources of intra-individual variation in biological or behavioral responses are of interest, multiple measurements per subject are evaluated (James, 1991). In these types of studies there is an inference that there is a differential in stressfulness of the different microenvironments; thus the stress response of the individual will change as the microenvironment experienced changes. These sorts of natural experiments on a population level are often conducted in ecological settings that are compartmentalized where the differences in sub-settings are clear and measurable. For example, the metropolitan New York City area (along with most other settings of urban sprawl) is an excellent setting for natural experiments designed to evaluate intra-individual stress responses (Harrison and Jefferies, 1977; James, 1991; Garruto *et al.*, 1999). The attractiveness of this setting for natural experimentation is that the social context, interpersonal interaction and physical environment will differ depending upon whether the study participant is engaged in economic activity (being at their place of employment), involved in leisure activity, or involved in domestic activity (being at their home). These different microenvironments are often quite separate from each other in terms of spatial distance, requiring significant time commuting from one to another (James, 1991).

Natural experiments that evaluate intra-individual variation also exist in migrant communities. Often, the migrant population lives in an enclave where, within that setting, indigenous sedente culture is maintained (Garruto *et al.*, 1999, Brown and James, 2000). The effects of exposure to the non-sedente culture (the culture of the group to which the participants have migrated) can be examined by contrasting stress-response measurements taken while participants are in the enclave with those taken while the participant is interfacing with the outside culture, such as while they are at their place of employment (Figure 9.3). Using this approach it is possible to make some evaluation of the stress response to culture change, at the individual subject level (Garruto *et al.*, 1999).

Model testing

From a modeling perspective, single-measurement studies assume that every subject would show the same biological response were it not for specific subject characteristics that cause the response to vary. The single-measurement approach is generally the basis for most

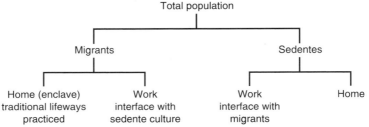

Figure 9.3. Natural experimental design highlighting the study of population enclaves.

traditional epidemiological studies of stress. It might be argued, however, that because stress implies either intra-individual change or ecologically based population differences studies that use single measurements and focus on between-subject differences within a population are not really studying stress per se. That is, stress response is inferred from the between-subject differences in characteristics, but not actually demonstrated. When there is only a single measurement per subject, there is potential confounding of the normal distribution of the single measurements (the inherent and non-stress related variation in the biological or psychological trait) with the response to stress. That is, a subject might have a given value simply because they are the 95th percentile of what might be a normal response, not because they are stressed.

There is also what might be termed "time motion" field studies of biobehavioral stress responses, in which multiple measurements are collected from each study subject without a predetermined framework other than the specifics of a particular population in a particular place. In these studies, the varying conditions extant at each measurement are of interest and are determined from diaries or direct observations recorded by either the subject or the researcher at the time of or immediately following each individual measurement (James, 2001). Data collected in this manner are often referred to as "ecological momentary data" (Schwartz and Stone, 1998; described in Chapter 3). In studies where this kind of data is collected, statistical modeling procedures are used to partition the variance in the biological or behavioral measurements associated with the recorded conditions and other characteristics of the participants in order to describe their effects. That is, effects are evaluated based on how well the data fit a statistical model.

The analysis of ecological momentary (EMA) data presents several challenges. There are measurements at multiple time points for each participant, and there is both within- and between-subject variation in the measurements. In a perfect world, there would be balanced data, meaning that there would be an equal number of measurements per potential stressor and potential stressor contrasts per subject. However, it is usually the case that the data are unbalanced, not only in terms of the number of measurements per subject, but also in terms of the stressor effect contrasts within subjects (factors which could influence the stress response that can be and are evaluated at each individual measurement). This problem, as discussed earlier, can be readily seen with ambulatory blood pressure measurements. Despite the fact that there is a pre-set sampling scheme (pressures taken every 15 minutes during the day), there is no control over the behavior of the individual. Hence the number of pressures taken during a given potential stressor will vary and the types of potential stressors will vary. Aside from ambulatory blood pressures and heart rates, ecological momentary data (stress responses) also include salivary cortisol measurements and serially collected psycho-metric assessments. Missing data are common with the EMA data when protocols require participant effort (e.g. saliva collection); there-fore, balanced data are particularly challenging to collect. To enable sufficient statistical power, investigators should over-sample to allow for missing data due to poor compliance and due to variance in mood experience over time.

What is of interest in studies of EMA data is the effects of the various factors (potential stressors) that were manifest at or during a stress-response measurement. However, because the data are unbalanced and unsystematic, a number of statistical problems arise with the data, which need to be addressed in order to obtain valid estimates of the effects of the potential stressors. A first problem is heteroscadasticity or a heterogeneity of variance in the data, both across the intra-subject factors (such as mood, location of measurement and posture) as well as between subjects. For example, the measurements taken during one circumstance (such as during sleep) are likely to be much less variable than measurements taken in another circumstance (while at work). Furthermore, a sample of measurements from any given subject is very likely to be different in terms of variance than that from any other given subject. There may also be relationships between a subject's sample mean and sample variance such that subjects with high mean values also have higher variance. This relationship is particularly true with regard to ambulatory blood pressure (Pickering, 1991). Variance may also change with time when there is a strong circadian pattern in the variable of interest (Ice, in press). There is also likely to be a problem with

the co-linearity (a close association) of between-subject effects
with subject-related covariates. For example, if there are multiple
measurements taken per subject and each subject has a different body
weight, it will be difficult to disentangle the "subject effect" from the
"body weight effect" on each individual stress response measurement
in the statistical modeling procedure because weight and subject are
essentially coterminous.

One potential way to avoid these issues in the analysis is to standardize
the data as noted in Chapter 6. Specifically, the mean value and
standard deviation for each subject are calculated, and each individual
measurement is then transformed to a z-score using the following
formula: $z_i = (X_i - \overline{X}_i)/s_i$ where:

> z_i = z-score of a given measurement for participant i;
> X_i = given measurement for participant i;
> $\overline{X}_i$ = participant i's overall mean; and
> s_i = participant i's overall standard deviation.

This procedure standardizes the total population of measurements,
adjusting the measurements so that participants whose intra-individual
biological or psychological variation is not very labile are as detectable
as participants whose measurements are labile (James *et al.*, 1986). In
addition, adjusting the values in this manner effectively eliminates the
between-subject effects in that all subjects have a mean of 0 and
a variance of 1. Because all subjects have the same mean, the effect of a
subject-associated covariate (such as weight) is no longer confounded
with the "subject effect." This approach has been used in several studies
of ambulatory blood-pressure responses to daily stress (James *et al.*, 1986;
James *et al.*, 1994; Brown *et al.*, 1998, Ice *et al.*, 2003).

Data analysis

Single-measure studies

Whether a natural experiment is conducted or a model-testing approach
is used, the analytic techniques to evaluate the within- and between-
subject variation in stress response are the same. Statistically, data from
single-measurement studies using either design approach would be
analyzed using analysis of variance/covariance (ANOVA/ANCOVA) or
regression analysis techniques. From a mathematical perspective, these
techniques are two sides of the same coin; that is, they are mathematically

equivalent (Neter and Wasserman, 1974; Green and Salkind, 2005). In using these techniques, a linear combination of the various independent parameters (potential stressors or participant characteristics) is formulated so that the variation associated with each can be estimated. A typical ANOVA/ANCOVA formulation would be:

$$Y_{ijkl} = \mu + \alpha_i + \beta_j + \alpha\beta_{ij} + \gamma_k + \varepsilon_{ijkl}$$

where:

Y_{ijkl} = the stress response measure on a given subject

μ = that portion of the stress-response measurement that is common to every subject (equivalent to the population mean)

α_i = the estimated variation associated with possible stressor A

β_j = the estimated variation associated with possible stressor B

$\alpha\beta_{ij}$ = the estimated variation associated with the interaction of possible stressors A and B

γ_k = the estimated variation associated with possible stressor C and

ε_{ijkl} = the remaining unexplained between subject variance in the stress response (random error)

The parameters can be a classificatory or "fixed" factor (nominal or ordinal variables, such as gender or urban/rural residence) or may be continuous or "random" (interval or ratio variables, such as age or weight). The determination of whether an effect is "fixed" or "random" is based on whether the alternative values of the effect define all the possibilities (fixed) or whether they are a sample of the possibilities (random). The decision as to whether a given parameter has a statistically significant effect on the stress response is based on the value of an adjusted ratio (F ratio) of the estimated variation associated with a possible stressor with the random error (Neter and Wasserman, 1974; Montgomery, 1976; Green and Salkind, 2005). It is critical to this analysis to define whether a parameter is "fixed" or "random" because that determination will affect the calculation of the F ratio, and thus the determination as to whether that parameter contributes significantly to the variation in the stress response (Montgomery, 1976). With a single-measurement study, the estimated variation is all "between subject," meaning that the effects examined evaluate why the stress response values vary among participants. In a natural experiment, stress responses are usually compared across "fixed" ecological contexts (such as sedente vs. migrant or modernized vs. traditional), whereas in a model framework, those responses may be compared across "fixed" personal characteristics of the study participant such as social class or ethnic group. Natural experiments might also include comparisons across "fixed" personal characteristics as well, so that the study design might evaluate gender and ethnic differences (personal characteristics) associated with sedente/migrant status (ecological context).

Using a regression approach, the same model would look like:

$$Y_i = \beta_0 + \beta_1 X_{1i} + \beta_2 X_{2i} + \beta_3 X_{1i} X_{2i} + \beta_4 X_{3i} + \varepsilon_i$$

where:

$Y_i =$ the stress-response measure on a given participant

$\beta_0 =$ that portion of the stress-response measurement that is common to every participant (equivalent to the population mean, but also the value where the line described by the regression model crosses the Y axis)

$\beta_1 =$ the estimated slope associated with possible stressor A (X_1)

$\beta_2 =$ the estimated slope associated with possible stressor B (X_2)

$\beta_3 =$ the estimated slope associated with the interaction between stressor A and B $(X_1 X_2)$

$\beta_4 =$ the estimated slope associated with possible stressor C (X_3)

and

$\varepsilon_i =$ the remaining unexplained variance in the stress response (random error)

The statistical tests in the regression analysis determine whether or not a given slope is significantly different from zero (Neter and Wasserman, 1974; Green and Salkind, 2005). If the slope is different from zero, the effect contributes to the variation in the stress response (Y_i). The "p value" or statistical significance associated with the tests of the slope of a possible stressor (X) will be identical to the F test of the same parameter in the ANOVA/ANCOVA model. The variation associated with the effects under both approaches is also the same. Thus, either analytic approach provides the same answer.

Multiple-measure studies

Studies that employ multiple measurements evaluate within-only or both within- and between-subject variation in the stress response. Depending upon whether a natural experimental approach or a model testing approach is used, the data will either be balanced (natural experiment) or unbalanced (ecological momentary data) within subjects. As previously noted, in a balanced study, each subject has the same number of measurements per potential stressor and there is the same number of measurements taken per participant. Studies using urine-collection techniques where urinary catecholamines or cortisol are evaluated (e.g. James *et al.*, 1985, 1993; Pollard *et al.*, 1996; Brown and James, 2000) and also some ambulatory blood-pressure studies (e.g. James *et al.*, 1993; James and Bovbjerg, 2001) have used balanced data.

The general form of the ANOVA/ANCOVA model that would be used to analyze balanced multiple measurements is:

$$Y_{ijklm} = \mu + \rho_i + \alpha_j + \beta_k + \alpha\beta_{jk} + \gamma_1 + \varepsilon_{ijklm}$$

where:

Y_{ijklm} = the stress-response measure on a given participant

μ = that portion of the stress-response measurement that is common to every participant (equivalent to the population mean)

ρ_i = the estimated variation across participants

α_{jk} = the estimated variation associated with possible stressor *A*

β_k = the estimated variation associated with possible stressor *B*

$\alpha\beta_{jk}$ = the estimated variation associated with the interaction of possible stressors *A* and *B*

γ_1 = the estimated variation associated with possible stressor *C*

and

ε_{ijklm} = the remaining unexplained between-participant variance in the stress response (random error)

This model is similar to the one for single-measurement studies, except the effect of person (subject) has been added. The subject effect is a "random" effect. In general, the type of design used to study balanced multiple measurements per participant is the "randomized complete block design" (Montgomery, 1976), where subjects are the "randomized blocks." These types of designs are also referred to as repeated-measures designs (Green and Salkind, 2005). Figure 9.4 illustrates the "randomized complete block design." As with single-measurement studies, if a regression approach was used, the modeling framework would mirror the ANOVA/ANCOVA approach, such that there would be an addition of person or subject effects. What would differentiate the regression

BLOCKS

Person 1	Person 2	Person 3	Person 4	....	Person N
Y_{ijklm1}	Y_{ijklm1}	Y_{ijklm1}	Y_{ijklm1}	....	Y_{ijklm1}
Y_{ijklm2}	Y_{ijklm2}	Y_{ijklm2}	Y_{ijklm2}	....	Y_{ijklm2}
Y_{ijklm3}	Y_{ijklm3}	Y_{ijklm3}	Y_{ijklm3}	....	Y_{ijklm3}
Y_{ijklm4}	Y_{ijklm4}	Y_{ijklm4}	Y_{ijklm4}	....	Y_{ijklm4}
.	.	.	.		.
.	.	.	.		.
.	.	.	.		.
Y_{ijklmn}	Y_{ijklmn}	Y_{ijklmn}	Y_{ijklmn}	....	Y_{ijklmn}

Figure 9.4.   The scheme for a randomized block design. The Y values are the measurements taken on each subject during each alternative for each potential stressor.

approach is that there would be as many participant parameters as there are subjects, as opposed to a single parameter per subject in the ANOVA/ANCOVA approach. What this means is that there would be a testable "slope" for each subject. In reality, the subject or person effects in the regression model are what might be termed "nuisance variables," in that they are designed to account for variation, but there is really no interest in testing their significance. Unfortunately, with many studies of stress markers, even if the study is designed to be balanced, poor compliance leads to unbalanced data. The investigator then has to choose to drop individuals with incomplete data to conduct ANOVA/ANCOVA analyses or to use the methods described below.

Finally, when using multiple measures that are unbalanced (ecological momentary data), the model employed will resemble that used for balanced data, except that analysis will be done using a "mixed effects" analysis. What differentiates this analysis from the randomized block design model is that the mixed-effects model uses a maximum likelihood algorithm to calculate the variance estimates for each of the potential stressor effects. The procedure does not require that all participants have each potential stressor in order for the various variance estimates to be made. It also does not require an equal number of measurements for each potential stressor. However, while the model can estimate effect sizes for all potential stressors, those estimates will depend upon the quality and quantity of data. The larger the data set, the better the estimates will be. An excellent example of the use of mixed models to evaluate stress-related ambulatory blood pressure changes is the study of traffic agents (people who hand out parking tickets) in New York City by Brondolo and colleagues (1999). Using mixed-effects models, they were able to show that both systolic and diastolic pressure varied by the social interaction and mood of the agents. Using a similar approach, van Eck *et al.* (1996) reported that cortisol increased with trait anxiety, agitation (state), negative affect (state) and stressful events. Using this modeling technique they were able to determine that the effect of agitation was random, suggesting that the impact of agitation varied by individual.

Statistical software

As previously noted, it is not the intent of this chapter to delve into specific statistical techniques. The statistical procedures noted above, however, are all available in the major statistical software packages such as SPSS, SAS and SysStat; thus to understand how they operate, the reader should consult the various reference manuals for the software. Finally, to conduct the analyses, all of these software packages use data organized into a spreadsheet, where each row defines the data for each individual subject and each column contains the different variables measured on each subject. Thus, in order to start the analysis process, data must be organized into a spreadsheet format. In many cases it is advantageous to organize data from different sources into separate files (e.g. interview, diary and physiological data) and then to merge the files prior to analysis. Relational databases (e.g. Access, Paradox, Filemaker Pro) can assist in file merging. Microsoft Access is also a useful tool for data entry because the "forms" page enables easier data entry than a traditional spreadsheet and data-entry rules can be set to disallow inappropriate values and thus minimize data-entry errors.

Data management, documentation and sharing

Data management is an often overlooked part of conducting research; however, poor data management can have significant consequences. Poor management can result in errors in data analysis and interpretation, interfere with data sharing and secondary data analysis and can lead to accusations of scientific misconduct in the event of an audit (Freedland and Carney, 1992). A key to good data management is proper data documentation. Data documentation, which is often contained in a three-ring binder, includes IRB and other research permission documents, information on how the sample was collected, how the data were collected, names and locations of the data files, contents of the data files, data manipulation and a master frequency distribution on all variables. Information on sample selection includes inclusion and exclusion criteria and documentation of unusual cases and reasons for dropouts. This information is required for IRB renewals. Data-collection information should include all measurement protocols, including interview protocols, codes assigned to data collectors and dates of data collection. If you've created a protocol book for training purposes, this will suffice for documentation with the addition of dates of data collection. Data files are often stored in more than one place. New data sets may be created for special analyses of subsets of the data or files may be used by more than one person, necessitating the creation of more than one file. This can lead to great confusion as to which file is the original data file and potentially could lead to data loss if the wrong file is erased (Freedland and Carney, 1992). Good documentation of the location of the files and subsequent manipulation of the files will avoid confusion. Several statistical packages now allow labeling of variables, descriptions of variables and code labels; however, many do not. A list of the variable names, codes and descriptions of the variables is good practice to ensure that analyses can be replicated. Furthermore, raw data are often used to create scales or indices. Unless the program to create those variables is saved, the scales cannot be re-created in the event of data loss or future studies. Documentation of the creation of such variables enables easy replication of such analyses. While variable frequencies and descriptive statistics can always be re-run, printing descriptive statistics and storing them with data documentation provides easier access as this information is needed for every new presentation or publication. Finally, it is also good practice to list the creation of subsets of data files for special analyses.

Another critical aspect of data management is data security. Proper storage of consent forms and identity of subjects is mandated by IRB. Paper copies of data forms should be locked and shred when no longer needed. Electronic copies should be password protected. Emailing data with identifiers should be avoided as email is not secure. A good back-up system is critical to prevent data loss.

Finally, most US funding agencies now require that data collected with grant money is available for data sharing. Furthermore, the National Science Foundation now requires that a data-sharing plan is submitted with grant proposals. Investigators are given ample time to conduct the analyses of interest but at some point all data need to be available to other scientists. Good data management and documentation are essential for data sharing.

What does it all mean?

The question that is most often asked when thinking about designing a stress study is which approaches are best? Should you conduct a natural experiment or simply test models? There is no simple answer to these questions. Stress research happens on a case-by-case basis, so that the best approach to use is the one that best fits the questions being addressed.

It should be realized that regardless of the type of approach used, the effects of stressors, which are determined from the statistical analysis of models, are estimates and not the perfect and true effect. Better estimates are obtained when there is both more and more complete data. Be aware that many of the stress effect estimates that can be calculated are also contingent on the statistical assumptions (for example, whether an effect is random or fixed) and computational approximations that are employed by the statistical program used. That is, the determination of whether or not a stressor has a biological effect may be determined not just by the evaluation of the experimental contrasts, but also by how you conceive of and how you define the potential stressors being tested. Hence, even with statistics, there is still a substantial subjective element in the analysis of the biological and psychological effects of stress.

References

Brondolo, G., Karlin, W., Alexander, K., Bubrow, A. and Schwartz, J. (1999). Workday communication and ambulatory blood pressure: implications for the reactivity hypothesis. *Psychophysiology*, **36**, 86–94.

Brown, D. E. (1981). General stress in anthropological fieldwork. *American Anthropologist*, **83**, 74–92.

 (1982). Physiological stress and culture change in a group of Filipino-Americans: a preliminary investigation. *Annals of Human Biology*, **9**, 553–63.

Brown, D. E. and James, G. D. (2000). Physiological stress responses in Filipino-American immigrant nurses: the effects of residence time, lifestyle and job strain. *Psychosomatic Medicine*, **62**, 394–400.

Brown, D. E., James, G. D. and Nordloh, L. (1998). Comparison of factors affecting daily variation of blood pressure in Filipino-American and Caucasian nurses in Hawaii. *American Journal of Physical Anthropology*, **106**, 373–83.

Brown, D. E., James, G. D., Nordloh, L. and Jones, A. A. (2003). Job strain and physiological stress responses in nurses and nurse's aides: predictors of daily blood pressure variability. *Blood Pressure Monitoring*, **8**, 237–42.

Dettenborn, L., James, G. D., van Berge-Landry, H. *et al.* (2005). Heightened cortisol responses to daily stress in working women at familial risk for breast cancer. *Biological Psychology*, **69**, 167–79.

Freedland, K. E. and Carney, R. M. (1992). Data management and accountability in behavioral and biomedical research. *American Psychologist*, **47**, 640–5.

Garruto, R. M., Little, M. A., James, G. D. and Brown, D. E. (1999). Natural experimental models: the global search for biomedical paradigms among traditional, modernizing and modern populations. *Proceedings of the National Academy of Sciences (USA)*, **96**, 10536–43.

Green, S. B. and Salkind, N. J. (2005). *Using SPSS for Windows and Macintosh: Analyzing and Understanding Data*, 4th edn, Upper Saddle River, NJ: Pearson Prentice Hall.

Harrison, G. A. and Jefferies, D. J. (1977). Human biology in urban environments: a review of research strategies. In *MAB Technical Note 3*, ed. P. T. Baker. Paris: UNESCO, pp. 65–82.

Ice, G. H. (in press). Factors influencing cortisol level and slope among community dwelling older adults in Minnesota. *Journal of Cross Cultural Gerontology*, In press.

Ice, G. H., James, G. D. and Crews, D. E. (2003). Blood pressure variation in the institutionalized elderly. *Collegium Antropologicum*, **27**, 47–55.

Jacob, R. G., Thayer, J. F., Manuck, S. B. *et al.* (1999). Ambulatory blood pressure responses and the circumplex model of mood: a 4-day study. *Psychosomatic Medicine*, **61**, 319–33.

James, G. D. (1991). Blood pressure response to the daily stressors of urban environments: methodology, basic concepts and significance. *Yearbook of Physical Anthropology*, **34**, 189–210.

(2001). Evaluation of journals, diaries, and indexes of worksite and environmental stress. In *Contemporary Cardiology: Blood Pressure Monitoring in Cardiovascular Medicine and Therapeutics*, ed. W. B. White. Totowa, NJ: Humana Press, pp. 29–44.

James, G. D. and Bovbjerg, D. H. (2001). Age and perceived stress independently influence daily blood pressure levels and variation among women employed in wage jobs. *American Journal of Human Biology*, **13**, 268–74.

James, G. D., Jenner, D. A., Harrison, G. A. and Baker, P. T. (1985). Differences in catecholamine excretion rates, blood pressure and lifestyle among young Western Samoan men. *Human Biology*, **57**, 635–47.

James, G. D., Pecker, M. S., Pickering, T. G. *et al.* (1994). Extreme changes in dietary sodium effect the daily variability and level of blood pressure in borderline hypertensive patients. *American Journal of Human Biology*, **6**, 283–91.

James, G. D., Schlussel, Y. R. and Pickering, T. G. (1993). The association between daily blood pressure and catecholamine variability in normotensive working women. *Psychosomatic Medicine*, **55**, 55–60.

James, G. D., Yee, L. S., Harshfield, G. A., Blank, S. and Pickering, T. G. (1986). The influence of happiness, anger and anxiety on the blood pressure of borderline hypertensives. *Psychosomatic Medicine*, **48**(6), 502–8.

McDade, T. W., Stallings, J. F. and Worthman, C. M. (2000). Culture change and stress in Western Samoan youth: Methodological issues in the cross-cultural study of stress and immune function. *American Journal of Human Biology*, **12**, 792–802.

McGarvey, S. T. and Baker, P. T. (1979). The effects of modernization and migration on Samoan blood pressures. *Human Biology*, **51**, 461–79.

McGarvey, S. T. and Schendel, D. E. (1986). Blood pressure of Samoans. In *The Changing Samoans: Behavior and Health in Transition*, ed. P. T. Baker, J. M. Hanna and T. S. Baker. New York: Oxford University Press, pp. 350–93.

Montgomery, D. C. (1976). *Design and Analysis of Experiments*. New York: Wiley.

Neter, J. and Wasserman, W. (1974). *Applied Linear Statistical Models*. Homewood, IL: Richard D. Irwin.

Pearson, J. D., James, G. D. and Brown, D. E. (1993). Stress and changing lifestyles in the Pacific: physiological stress responses of Samoans in rural and urban settings. *American Journal of Human Biology*, **5**, 49–60.

Pickering, T. G. (1991). *Ambulatory Monitoring and Blood Pressure Variability*. London: Science.

Pollard, T. M., Ungpakorn, G. and Harrison, G. A. (1996). Epinephrine and cortisol responses to work: a test of the models of Frankenhaeuser and Karasek. *Annals of Behavioral Medicine*, **18**, 229–37.

Schwartz, G. E. Weinberg, D. A. and Singer, J. A. (1981). Cardiovascular differentiation of happiness, sadness, anger and fear following imagery and exercise. *Psychosomatic Medicine*, **43**, 343–64.

Schwartz, J. G. and Stone, A. A. (1998). Strategies for analyzing ecological momentary assessment data. *Health Psychology*, **17**, 6–16.

Schwartz, J. E., Warren, K. and Pickering, T. G. (1994). Mood, location and physical position as predictors of ambulatory blood pressure and heart rate: Application of a multilevel random effects model. *Annals of Behavioral Medicine*, **16**, 210–20.

Sterling, P. (2004). Principles of allostasis: optimal design, predictive regulation, pathophysiology, and rational therapeutics. In *Allostasis, Homeostasis, and the Costs of Physiological Adaptation*, ed. J. Schulkin. New York: Cambridge University Press, pp. 17–64.

Van Eck, M., Berkhof, H., Nicolson, N. and Sulon, J. (1996). The effects of perceived stress, traits, mood states, and stressful daily events on salivary cortisol. *Psychosomatic Medicine*, **58**, 447–58.

10 Protection of human subjects in stress research: an investigator's guide to the process

GARY D. JAMES AND GILLIAN H. ICE

All studies involving human subjects must be reviewed and approved by an Internal Review Board (IRB). This board consists of both professional and non-professional (lay) members, and its mission is to reasonably insure that subjects in research are adequately informed about the procedures that they will undergo and that they are protected from harm. In order for successful human stress research to proceed, an investigator needs not only to understand what constitutes ethical research, but also how to convey to the IRB that best practices will be followed in the conduct of the research and that any risks to subjects are minimized. Furthermore, if there are risks to subjects, these must be explained plainly and carefully to the subjects in a consent form prior to their participation, so that the subjects can make an informed decision as to whether or not they want to proceed. Finally, if there are risks, it must be made clear that they are substantially outweighed by the potential benefits to the subject or society. In fact, it is not possible to obtain federal grant money in the USA to study stress or allostatic processes in humans without documenting the protection of human subjects and IRB approval.

Researchers often view the entire process of assembling and submitting human subjects review forms as an onerous and difficult task. Perhaps the single major complaint (or frustration) that is voiced by researchers is that they do not understand what the IRB wants. Many researchers in human biology and biological anthropology receive little if any training with regard to ethics in research, which contributes to their difficulty with the human subjects review process. The purpose of the chapter is to outline the process of IRB review and approval for researchers interested in studying the effects of stress in humans. Investigators should realize, however, that it is not enough to simply write a good IRB protocol. The actual conduct of the research must follow to the letter the procedures outlined in the IRB protocol. To willfully mislead the IRB by detailing one set of procedures but then doing another in the conduct of the research has dire consequences for both the subjects and the investigators.

Guidelines for ethical research using human subjects

There are any number of examples of abuses in human subjects research since its inception extending back into Greek and Roman antiquity (Vanderpool, 1996), and a great deal has been written about them. While it is useful to know about these lapses in ethics, it is not the intent of this chapter to chronicle man's inhumanity to man. Detailed descriptions of various ethical problems in human research (up to the present day) can be found in documents on the websites of the National Institutes of Health's Office of Human Subjects Research (OHSR) (http://ohsr.od.nih.gov/) and the CITI Course in the Protection of Human Research Subjects (http://www.citiprogram.org/).

The modern codification of human subjects research ethics began in 1947 with the Nuremberg Code, which was developed as a result of the atrocities against humans during World War II by Nazi researchers. There were ten points to the Code, including: subject consent must be voluntary and freely given; that the benefits of the research must outweigh risks; and that participants maintain the ability to terminate participation at any time. In 1964, the World Medical Association developed a code of research ethics that came to be known as the Declaration of Helsinki, which elaborated on the principles in the Nuremberg Code with an eye toward medical research with therapeutic intent. The Declaration has been revised several times over the years (1975, 1983, 1989, 1996, 2000). A copy of the declaration can be found at the OHSR website (http://ohsr.od.nih.gov/). Many journal editors require that research be performed in accordance with this Declaration. In principle, this document sets the stage for the implementation of the IRB process (http://www.citiprogram.org/).

In 1974, the United States Congress authorized the formation of the National Commission for the Protection of Human Subjects in Biomedical and Behavioral Research, which has come to be known as The National Commission (http://www.citiprogram.org/). Congress charged the National Commission to identify the basic ethical principles that underlie the conduct of human research, and also asked it to develop guidelines to assure that human research is conducted in accordance with those principles. In 1979, The National Commission published the Belmont Report, which is required reading for anyone conducting human subjects research. The Report identifies three basic ethical principles that underlie all human subjects research, commonly called the Belmont Principles, which are respect for persons, beneficence, and justice. A copy of the report can be found at the OHSR website (http://ohsr.od.nih.gov/).

Finally, in the 1970s the US Congress formed an Ad Hoc Panel that recommended that federal regulations be designed and implemented to protect human research subjects in the future. Subsequently, federal regulations were enacted including the National Research Act, 45 46 and 21 Code of Federal Regulations 50 (http://www.citiprogram.org/). These regulations outline the guidelines for human subjects research in the USA today, defining the structure and functions of institutional IRBs. A copy of the regulations can be found at the OHSR website (http://ohsr.od.nih.gov/). The Office of Human Research Protection (OHRP) and the Food and Drug Administration (FDA) are the agencies of the federal government that oversee the conduct of institutional IRBs. These federal agencies have the authority to stop all research at an institution if they believe there is evidence of inadequate protection of human research subjects. Similar human subjects laws have been enacted in most countries around the world, based on the Declaration of Helsinki. While we focus on the US rules and process in this chapter, researchers should educate themselves in the laws of the country in which they work. Researchers are bound by the rules which govern the treatment of human subjects both in their country of origin and the country in which they conduct the research.

IRB structure

Researchers should not view the IRB as being composed of their peers. The committee is really not evaluating the scientific merits of the investigator's research per se. Rather, its focus is on assuring that the subjects in the research are fully informed about the risk of their participation and that the risk of harm to the subjects related to study participation is minimized. When there is substantial risk to the participant, the committee will weigh the risk of participation against the benefits of the research to the subject and society in determining whether it can be undertaken.

Title 45, Section 46 of the U.S. Federal Code of Regulations defines the minimum size and make-up of the IRB. It stipulates that each Internal Review Board (IRB) shall have at least five members. In practice, most IRBs are substantially larger, owing to the need for diverse scientific expertise and advocacy for special populations who are likely to be studied by researchers at an institution.

The Federal Code also prescribes the nature of the membership: specifically, no IRB shall consist entirely of men, entirely of women or entirely of members of one profession, each IRB shall include at least one member whose primary concerns are in nonscientific areas, and at least one member who is not otherwise affiliated with the Public Health Service. Finally, the Code also allows the IRB chair to appoint ad hoc members with competence in special areas to assist in the review of protocols, but these ad hoc members may not vote to approve or disapprove the protocol. This last point is important, in that researchers often complain that the committee does not understand their research because there is no member of the committee who has expertise in their area. Most institutions will post the membership of the committee on their website, or otherwise provide a listing of the membership. If you feel that the committee membership does not have the appropriate expertise, you could request that ad hoc members be appointed when your protocol is reviewed, although it is likely that the Chair will do it before your request.

Determining factors in IRB research review: What are they looking for?

Most IRB Human Subjects Research Forms have several standard items that must be filled out. In filling out the form for IRB review, the best policy is to write in the plainest, most straightforward language possible. Avoid jargon at all costs. There is always a research summary that must be in everyday language, because as previously noted there are members of the IRB who are non-professionals. Expect that somewhere between one quarter to one half of the committee will be non-scientists, and that a goodly number of the remaining committee will be scientists from another discipline. If these non-expert members are confused by your description of the study, it will take longer to get approval.

When the committee evaluates the protocol, they are examining several factors, each of which is addressed in various items in the human subjects protocol form. For the committee to approve the protocol, the researchers must provide sufficient information so that the following questions can all be answered "yes" (from OHRP IRB Guidebook, http://www.hhs.gov/ohrp/irb/irbguidebook.htm).

1. Are the risks to subjects minimized or reasonable in relation to anticipated benefits?
2. Is the selection of subjects equitable?
3. Is informed consent sought?
4. Is informed consent appropriately documented?
5. Are there adequate provisions for the monitoring of data collection to ensure the safety of subjects (that is, are the measurement procedures safe)?
6. Are there adequate provisions to protect the privacy of subjects and confidentiality of the data?
7. If the study is taking place in another institution or overseas, has approval of other required groups been sought?
8. If the study is taking place in a foreign country, has the study been reviewed and approved by a local human subjects committee?

Confidentiality and risk

One of the most important risks in biomedical and biosocial research is the effect of a breach of confidentiality. The researcher must make every effort to try to ensure confidentiality; we suggest relying on anonymity whenever possible. Anonymity means that no one, not even the investigator, can identify an individual subject. Simply eliminating names and other obvious identifiers does not guarantee anonymity; demographics can sometimes identify subjects as well. Any information or pattern of information that can uniquely identify an individual eliminates anonymity. When anonymity is not feasible, then the researcher must demonstrate to the IRB how confidentiality can be assured. Depending on the sensitivity of the subject matter, extra care should be taken to ensure that subjects cannot be identified. With some research topics, such as sexuality or criminality research, not only is the information sensitive, but the subjects' presence in the study can be a sensitive piece of information. Researchers must also be aware that demographic information, in some circumstances, can be an identifier. If the research concerns illegal behavior (e.g. a study of HIV and risk factors among prostitutes) the researcher may need to have the cooperation of local legal authorities or a federal Certificate of Confidentiality (see below). If there is a risk of triggering retribution by others, such as violence by abusing partners, the researcher must ensure that nothing given can identify a person as a respondent. Finally, risk to the community must also be minimized. This can be accomplished by having the researcher and community agree about publication (e.g., whether to identify the community) prior to the initiation of the research.

Certificate of Confidentiality

US researchers can apply for a Certificate of Confidentiality through the United States Assistant Secretary for Health that will provide protection even against a subpoena for research data. The policy defines sensitive research as involving the collection of information falling into any of the following categories: (a) information relating to sexual attitudes, preferences, or practices; (b) information relating to the use of alcohol, drugs, or other addictive products; (c) information pertaining to illegal conduct; (d) information that if released could reasonably be damaging to an individual's financial standing, employability, or reputation within the community; (e) information that would normally be recorded in a patient's medical record, and the disclosure of which could reasonably lead to social stigmatization or discrimination; and (f) information pertaining to an individual's psychological well-being or mental health (from OHRP IRB Guidebook, http://www.hhs.gov/ohrp/irb/irbguide-book.htm). It is important to note that you do not have to have federally funded research to obtain a Certificate of Confidentiality.

Study approval: expedited and full-board review

Study approval is obtained from either an expedited or full-board review. For full-board reviews approval is obtained by a majority vote of the IRB, whereas in an expedited review the chair or chair designate approves the protocol, acting on behalf of the full board. Expedited review, by definition, should be quicker, although some institutions have administrative procedures that make an expedited review as long or longer than a full-board review, from the viewpoint of the researcher.

A study protocol can be expedited if the procedures are carried out through standard methods and when there is no more than minimal risk to the study subjects. Minimal risk procedures are those where "the probability and magnitude of harm or discomfort anticipated in the research are not greater in and of themselves than those ordinarily encountered in daily life or during the performance of routine physical or psychological examinations or tests." (Title 45 CFR Part 46, 2001, 102 (i)). In 1998, guidelines defining these procedures and to whom they applied were issued by the US Secretary for Health and Human Services (Federal Register, 1998). As listed in 63 FR 60364–60367, November 9, 1998. These were subsequently updated when Title 45, CFR Part 46 was updated in 2001 and again in 2005. Procedures that can be expedited are as follows:

(a) Research conducted in established, or commonly accepted educational settings, involving normal educational practices, such as (i) research on regular and special educational instructional strategies, or (ii) research on the effectiveness of or the comparison among instructional techniques, curricula, or classroom management methods.

(b) Research involving the use of educational tests (cognitive, diagnostic, aptitude, and achievement), if the information taken from these sources is recorded in such a manner that subjects cannot be identified, directly or through identifiers linked to the subjects.

(c) Collection of hair and nail clippings, in a non-disfiguring manner, deciduous teeth and permanent teeth, if patient care indicates a need for extraction.

(d) Collection of excreta and external secretion including sweat, uncannulated saliva, placenta removed at delivery, amniotic fluid at time of rupture of the membrane prior to or during labor.

(e) Recording of data using noninvasive procedures routinely employed in clinical practice. This includes the use of physical sensors that are applied either to the surface of the body or at a distance and do not involve input of matter or significant amounts of energy into the subject or an invasion of the subject's privacy. It also includes such procedures as weighing, testing sensory acuity, electrocardiography, electroencephalography, thermography, detection of naturally occurring radioactivity, diagnostic echography, and electroretinography. It does not include exposure to electromagnetic radiation outside the visible range (for example, x-rays, microwaves).

(f) Collection of blood samples by venipuncture, in amounts not exceeding 450 milliliters in an eight-week period and no more often than two times per week, from subjects 18 years of age or older and who are in good health and not pregnant.

(g) Collection of both supra and subgingival dental plaque and calculus, provided the procedure is not more invasive than routine prophylactic scaling of the teeth and the process is accomplished in accordance with accepted prophylactic techniques.

(h) Voice recordings made for research purposes, such as investigations of speech defects.

(i) Moderate exercise by healthy volunteers.

(j) The study of existing data, documents, records, pathological specimens, or diagnostic specimens.

(k) Research on individual or group behavior characteristics or individuals, such as studies of perception, cognition, game theory or test development, where the investigator does not manipulate subjects' behavior and the research will not involve stress or deception of subject.

(l) Research on drugs or devices for which an investigational new drug exemption or an investigational device exemption is not required.

(Federal Register, 1998; Title 45 CFR Part 46, 2005)

The categories in this list apply regardless of the age of subjects, except as noted. It should be remembered that these are guidelines, so that even if they apply, the IRB may decide to give the protocol a full-board review anyway, due to other factors. Such factors may be where identification of the subjects and/or their responses would reasonably place the subjects at risk of criminal or civil liability or be damaging to the subject's financial standing, employability, insurability, or reputation, or be stigmatizing (Federal Register, 1998). The procedures of field stress research are mostly defined as minimal risk (collection of body fluids, taking blood pressure, collecting questionnaires). Finally, if the investigator wants to change any procedure that will affect the study subjects, approval for the changes from the IRB must be obtained.

Certain research participants are considered to be vulnerable populations. Research with these participants will often require full-board review regardless of protocol. Vulnerable populations include those with developmental disabilities or who are cognitively impaired, children and prisoners. Principle investigators may have to describe special procedures to protect these subjects.

The IRB is also required by US federal regulations to re-evaluate research protocols at intervals appropriate to the degree of risk to the study subjects, with the maximal time between reviews being one year. For research involving no more than minimal risk (which would generally include field studies of stress), the approval period is one year.

For research involving greater than minimal risk, the IRB determines the approval period. Investigators must apply for continuation of approval if the study exceeds the period of IRB re-evaluation. Continuing review by the IRB of a protocol usually occurs for three to five years, the specified period being determined by the IRB. After the period, the protocol must be resubmitted for regular review.

Continuing review forms usually require investigators to provide the following information (from http://binghamton.edu/ipph/compliance1. html):

- confirmation that the procedures have not changed
- the number of subjects taking part to date
- a description of any adverse events or unanticipated problems involving human subjects that occurred since the last review
- the number of subjects who withdrew from the research, and any complaints about the research they had
- a description of preliminary findings
- a copy of the current informed consent document and/or all information provided to subjects
- any and all changes in personnel

Continuing review will occur on or before the anniversary date of the previous IRB review, even though the research activity may not begin until some time after the IRB has given approval. The IRB will usually send a notification to the investigator one to two months prior to the anniversary date. It is important to note that if approval for the protocol lapses continued conduct of the research is a violation of federal regulations and could lead to a variety of sanctions against the researcher.

Finally, US federal regulations (Title 45 CFR Part 46, 2005, Subpart A) require that researchers report "adverse events" to the IRB. An adverse event is an occurrence or situation during the course of a research project that includes any serious event associated with the study procedures and/or problems involving the conduct of the study which may occur during the course of the researcher's own research projects. Any problems involving the conduct of the study or patient participation must be reported. For example, social and behavioral interviews may deal with sensitive issues – occasionally, research subjects will become upset because of the nature of the questions and this requires reporting. Any problems involving the recruitment and/or consent processes require reporting. For example, if a person who is contacted, either in writing or in person, about participating in a study becomes upset about the recruitment process, this should be reported. Any deviations from the approved protocol should be reported in writing. Examples of a more serious nature include incidents of a person being enrolled in a study before signed consent has been obtained. All adverse events must be reported promptly.

The IRB will review the adverse event reports in order to re-evaluate the risks/benefit of the study and/or the appropriateness of the recruitment/consent process to determine if any changes should be made in the protocol or consent form. If the investigator has already modified the protocol or consent form in response to these events, the appropriateness of these changes is also reviewed. Failure to report is a breach of the conditions under which IRB approval is given, and could result in suspension or revocation of approval. Suspension or revocation of approval could result in loss of support by funding agencies and loss of right to publish.

Getting informed consent

The consenting process in adults

Informed consent is one of the primary ethical principles governing human subjects research; it assures that prospective human subjects will understand the nature of the research and can knowledgeably and voluntarily decide whether or not to participate. "Informed consent" means the knowing consent of an individual, or his/her legally authorized representative, who is able to exercise free power of choice without undue inducement or any form of force, fraud, deceit, duress or other form of constraint or coercion. Documentation of "legally effective informed consent" usually involves the use of a written consent form containing all of the information to be disclosed and signed by the subject or the subject's legal representative. It should be emphasized that the consent form is merely the documentation of informed consent and does not, in and of itself, constitute informed consent. The fact that a subject signed a consent form does not mean that he/she understood what was being agreed to or truly gave their voluntary consent. Informed consent is a process that is documented by a signed consent form (from http://www.binghamton.edu/ipph/compliance1.html).

In order to be valid, consent must be freely given, meaning that it is free from all forms of coercion. In addition to overt coercion, the investigator needs to be sensitive to more subtle forms of coercion, such as social pressure, requests from authority figures, and undue incentive for participation (e.g. excessive amounts of money for participation). Since coercion exists when it is perceived by the subject, the investigator must attempt to view the consent process from the subject's perspective. For example, a professor asks his/her students to participate in research. Even if he/she tells the students that participation will not affect their grades, most students will assume that they will somehow be penalized for not participating. A further complication exists in therapeutic and educational settings. People often assume that medical personnel and teachers are acting in their best interest. Evidence has indicated that this assumption persists even if the subjects are told that the activity is research and will have no direct benefit for them. Therefore, special care must be taken in these settings to ensure that the potential subjects understand the nature of the research.

Informed consent is not a single event or just a form to be signed — rather, it is an educational process that takes place between the investigator and the prospective subject. The basic elements of the consent process include full disclosure of the nature of the research and the subject's role, adequate comprehension on the part of the potential subjects, and the subject's voluntary choice to take part. US federal regulations (Title 45 CFR Part 46, 2005) detail the following elements of information that must be provided to each subject:

- A statement that the study involves research, an explanation of the purpose of the research, the expected duration of the subject's participation and a description of the procedures to be followed;
- A description of any foreseeable risks or discomforts to the subject;
- A description of any benefits to the subject or to others which may be reasonably expected from the research;
- A statement describing the extent to which confidentiality of records identifying the subject will be maintained;
- An explanation of who to contact for answers to pertinent questions about the research and research subjects' rights, and who to contact in the event of a research-related injury;

- A statement that participation is voluntary, that refusal to participate will involve no penalty or loss of benefits to which the subject is otherwise entitled, and that the subject may discontinue participation at any time; and
- For research involving more than minimal risk, an explanation that your institution does not have a formal plan or program to provide medical treatment or compensation for any injury which occurs as a result of the subject's participation (the subject should also be informed that this does not waive any of his/her legal rights).
- The Human Rights Statement: "if at any time you have questions concerning your rights as a research subject you may call the Chair of the IRB at ..."

When appropriate, one or more of the following elements of information should also be provided to each subject:

- a disclosure of appropriate alternative procedures or courses of treatment, if any, that might be advantageous to the subject;
- an identification of any procedures which are experimental;
- a statement that the research may involve risks to the subject which are currently unforeseeable;
- anticipated circumstances under which the subject's participation may be terminated without regard to the subject's consent;
- any additional costs to the subject that may result from participation;
- the consequences of a subject's decision to withdraw from the research and procedures for orderly termination of participation by the subject;
- a statement that significant new findings developed during the course of the research which may relate to the subject's willingness to participate will be provided to the subject; and
- the appropriate number of subjects in the study.

Consent forms should be designed to meet the needs of the particular research project where it is being used; no one form can be used in every research project. Good consent forms meet the following four criteria (from http://www.binghamton.edu/ipph/compliance1. html).

1. They are brief, but have complete basic information. Many potential subjects do not read long consent forms. The longer the form, the smaller the number of people who read it in its entirety, and the smaller the fraction of it that is read by the rest. That is, the quest to be more comprehensive by including more information may actually result in the information transmitted being less comprehensive.

Include only the basic information needed by potential subjects ("basic" are the items required by federal regulations) and do not try to answer every conceivable question. "Non-basic information" can be given in a separate handout, perhaps in a question-and-answer format. One suggestion is to include a list of questions at the beginning of the handout, to permit each person to go to those questions that most interest him or her.

2. They are readable and understandable to most people. Articles in most popular magazines are at the 8th-grade level. Several computer programs estimate readability by the Flesch, Flesch–Kincaid, and FOG measures. Factors that improve readability include 1) technical terms should be replaced with ordinary language; 2) use active tense rather than passive tense verbs ("we did" rather than "it was done"); 3) write shorter sentences in general; and 4) make clear the links of logical sequences and of cause-and-effect, even if doing so makes the sentence much longer ("we will do this, because that happened").

3. They are in a format that helps people comprehend and remember the information. Format can be used to help people comprehend and remember complex material. Good format uses are: headings, indents, bolded type, lists, extra spacing between sub-topics, repetition, reasonable-size type, and plenty of margins and empty space in general. Those formats help the reader to organize the information, recognize, know, and remember the key points and go back later to the consent form and retrieve important information (such as telephone number of the investigator to call with questions).

4. They serve as a script for the face-to-face discussions with the potential subjects. Face-to-face discussions between researcher and potential subject are the most important part of the process of informed consent. These sample forms can be the script for the verbal explanation by the researcher. If the verbal explanation is almost identical to the written consent form, each will reinforce the other and potential inconsistencies will be avoided. One benefit of this approach is that the form or script prompts the researcher to use simple language for the verbal explanation. Another benefit is that the same form or script can be used for potential subjects who have difficulty reading or low literacy or who need a translation. This should also improve consistency of explanation among all subjects. Researchers need develop only one form or script, not two, to permit people of all literacy levels to be potential subjects. The script could

also be used in videotaping the explanation. On the other hand, it is not advisable simply to read the consent form to participants — it must be explained!

Finally, informed consent is not valid unless the study subject understands the information that has been provided. While no one can guarantee that another person has understood the information presented, it is the responsibility of the investigator to do what he/she can to enhance each prospective subject's comprehension of the information. To determine the appropriate way to present the information, the investigator must consider: the characteristics of the proposed subject population; the type of information to be conveyed; and the circumstances under which the consent process will take place (e.g., manner, timing, place, personnel involved). Factors such as age, education level, cognitive ability, and language fluency directly affect subject comprehension of information. In most cases, technical terms and complex sentences should be avoided, even for the educated layperson. Technical terms should be replaced with ordinary language, and short sentences using active tense rather than passive tense verbs should be used. If English is not the subject population's primary language, the explanations and forms may need to be translated and back translated (to ensure accurate translation). For children, care must be taken to ensure that the language is appropriate to their age level. For elderly subjects, oral information may have to be presented slowly and loudly and forms printed in large type. When the subject population is not homogeneous, different consent procedures may be required with different populations. The investigator should be aware that, even if the IRB has approved a consent procedure, it is his/her responsibility to ensure that each potential subject understands the information and to take whatever steps are necessary to gain that comprehension (from http://www.binghamton.edu/ipph/compliance1.html).

Consenting process in minors: parental permission

In studying children, it is often the case that the researcher must obtain parental permission as well. Documentation of parental permission depends on the nature of the research. In most cases, a signed form should document parental permission. On this form, parents must be provided with the following information:

1. a statement that the study is a research project;

2. identification of the investigator;
3. a statement of the purpose of the study;
4. an explanation of how and why their child was selected;
5. an explanation of the procedures, including setting, time involved, and with whom the child will be interacting;
6. a description of any discomforts or risks;
7. a description of any benefits to the child;
8. a clear indication that participation is voluntary and confidential, and that subjects may withdraw at any time without penalty;
9. an offer to answer questions along with information for contacting the investigator involved, and the human rights statement, "if at any time you have questions concerning your rights as a research subject you may call the Chair of the IRB at...";
10. they do not have to answer any question(s) they do not want to (questionnaires, survey studies) (from http://www.binghamton.edu/ipph/compliance1. html).

In addition, parents must be informed as to what procedures will be followed for subjects whose parents have not given permission for them to participate. All information must be in language that is understandable to the parent. For research conducted in schools, it is essential that clear indication be given that this research is separate from regular school activity and that it is not being conducted under the auspices of the school.

Assent

In addition to parental permission, adequate provisions must be made for soliciting the assent of the children capable of providing it. Assent means that the child gives affirmative agreement to participate in the research. Failure to object on the part of the child should not be construed to mean that they assent.

Procedures for soliciting the assent of children must be appropriate for the age level, maturity and psychological state of the child, and the IRB decides the age of a child when assent needs to be obtained. The essential information given to the child must include a description of the procedures and clear indication as to the voluntariness of his/her participation. If a child cannot give verbal assent observational measures should be used. Indicators might include a child physically resisting a procedure (such as a child who refuses to open his/her mouth for saliva collection) or crying which appears to be a direct result of research procedures. Investigators should clearly write such an assent protocol and clearly communicate it to all research staff. As with parental permission, when research is conducted in schools, a clear indication must also be given that this research is not part of the child's regular school program, is not being conducted under the auspices of the school, and the child's grade will not be affected by his/her decision to participate. This information must be presented in language that is understandable to the child. In developing language appropriate for the child in a given study, it is recommended that investigators unfamiliar with the capabilities of the population being studied consult with experts. For the researcher, in cases where there is inconsistency between the permission of the parent and the assent of the child, the following rule should be followed: a "no" from the child overrides a "yes" from the parent, but a "yes" from the child does not override a "no" from the parent.

In addition to requiring assent from children, investigators must develop a protocol for establishing consent and assent in adults with developmental disabilities or cognitive impairments. These participants are often unable to give truly informed consent. When working with such populations, researchers should seek permission from a family member or legal guardian. Older adults with dementia often do not have an appointed legal guardian; however, care providers can typically provide the researcher with the name of a family member who is responsible for medical decision making. Adults with cognitive impairments often cannot communicate yes or no verbally so researchers should develop a protocol which includes behavioral cues for refusal of assent, for example, if a participant repeatedly removes an ambulatory blood-pressure cuff or refuses to open his/her mouth for saliva collection. Similar to assent in children, a "no" from a cognitively impaired or developmentally disabled adult overrides a "yes" from a legal guardian but a "yes" from the impaired adult does not override the guardian's "no" (from http://www.binghamton.edu/ipph/compliance1. html).

Research involving deception

Since there is no informed consent in research involving deceptions, the terms "Informed Consent" or "Consent Form" should not be used. However, subjects still need to be informed about the nature of the research in a way that does not invalidate the data. One possibility is to provide subjects with a "Subject Information Sheet" which contains the necessary information. Subjects must be provided with sufficient information for them to decide whether to participate and, as in all other human subjects research, be allowed to withdraw at any point without penalty.

All research involving deception must be reviewed by the full IRB. Deception cannot be used in any study where there is significant risk to subjects. Finally, all subjects must be debriefed regarding the true nature of the research after their participation.The debriefing should: 1) explain all truths not revealed and all falsehoods told to subjects; 2) address the reasons the deceptions were necessary and 3) reassure subjects that their reactions to the deceptions were normal.

If having incomplete or erroneous information is not likely to be harmful to subjects, the IRB may consider allowing the researcher to delay the debriefing until all subjects have completed their participation. Care should be taken in debriefing to protect the well-being of the subjects. A written text of the explanation must be submitted to the IRB. This text can serve as the basis for conducting the debriefing; however, debriefing should always be a dialog. Those conducting the debriefing should be trained to elicit and respond to subject concerns. The use of highly evaluative terms (e.g. "We tricked you" or "We lied to you") should be avoided in explaining deceptions to the subjects (from http://www.binghamton.edu/ipph/compliance1. html).

Waived written informed consent

IRBs may approve a consent procedure that does not include, or which alters, some or all of the elements of written informed consent, or waive the requirements to obtain written informed consent provided that the investigator documents that: 1) the research involves no more than minimal risk to the subjects; 2) the waiver or alteration will not adversely affect the rights and welfare of the subjects; 3) the research could not practicably be carried out without the waiver or alteration; and 4) whenever appropriate, the subjects will be provided with additional information after participation (from http://www.binghamton.edu/ipph/compliance1. html).

When written consent is waived, researchers must still ensure that subjects are giving informed consent to participating in the study. Often, the consent may take some form of cover sheet or instructions to subjects that provide them with the same information that would be included in a consent form.

Investigators may choose to get a waiver of a signed consent form for research participants who are illiterate. With such participants, the request to sign a document which they cannot read can be interpreted as placing a participant at risk. IRBs at different institutions and in different countries interpret the level of risk for illiterate populations differently. As an alternative a witness can often sign on the participant's behalf.

Health insurance portability and accountability act regulations: getting permissions

In 1996, the US Congress passed the "Health Insurance Portability and Accountability Act" (HIPAA for short), which was put into effect April 14, 2003. The Act governs the privacy of personal protected health information (PHI). Researchers should be aware of HIPAA because it establishes the conditions under which covered entities can disclose PHI for research.

HIPAA does not apply to research per se; it apples to entities (such as hospitals, universities or other institutions) that collect and house PHI. The Act affects researchers because it affects their access to this information. Table 10.1 lists the 18 elements that could be used to identify the health information of an individual or the individuals' relatives, employers, or household members (from 45 CFR Parts 160 and 164; Standards for Privacy of Individually Identifiable Health Information, 2002).

HIPAA has provisions that allow covered entities to disclose PHI for research. In addition, in certain circumstances, covered entities may allow researchers to use PHI without subject authorization for certain types of research activities. Basically, however, a researcher can use medical data if all identifiers (see Table 10.1) are removed.

When a researcher needs to identify the data that they are collecting from the medical records, the researcher can also have patients (subjects) fill out a HIPAA permission form, or a Privacy Rule Authorization form. This form is separate from the consent form, although some institutions allow a merger of the two. A valid Privacy Rule Authorization is an individual's signed permission that allows a covered entity to disclose the individual's PHI for the purposes stated in the authorization. There is specific language that the authorization must have, but because it is the entity and not the researcher who is responsible for the PHI, the entity will have forms that the researcher must use to access PHI of study subjects. If PHI is needed for a research project, all rationale and, if necessary, the authorization form needs to be included in the IRB submission for the project.

HIPAA was not intended to impede research. Rather, from a research perspective, HIPAA describes methods to de-identify health information

Table 10.1. *The list of 18 elements that could be used to identify the individual or the individuals' relatives, employers, or household members from medical records.*

1. Names
2. All geographic subdivisions smaller than a state, including street address, city, county, precinct, ZIP code, and their equivalent
3. All elements of dates (except year) for dates directly related to an individual, including birth date, admission date, discharge date, date of death; and all ages over 89 and all elements of dates indicative of such age
4. Telephone numbers
5. Facsimile numbers
6. Electronic mail addresses
7. Social Security numbers
8. Medical record numbers
9. Health plan beneficiary numbers
10. Certificate/license numbers
11. Account numbers
12. Vehicle identifiers, serial numbers including license plate numbers
13. Device identifiers and serial numbers
14. Web universal resource locators (URLs)
15. Internet Protocol (IP) address numbers
16. Biometric identifiers, including fingerprints and voiceprints
17. Full-face photographic images and any comparable images
18. Any other unique identifying number, characteristic, or code

Source: 45 CFR 164.514.

such that it is no longer governed by the Act. If de-identification is not possible, covered entities (and hence researchers) can obtain the individual's written permission for access to their PHI in an authorization document describing the research uses of the PHI and the rights of the research subject. When obtaining authorization is not practicable, the entity's Privacy Board (which is often the IRB) can also waive or alter the authorization requirement, but this is usually only done when it is apparent that there is minimal risk to the study subjects.

Researcher educational requirements

Prior to the late 1990s, when ethical problems occurred in human subjects research, institutional IRBs were punished. However, it became clear that when researchers deceived the IRBs, the IRBs have little control over the conduct of research. Attention then focused on the researchers, who, in cases where subjects were harmed, often argued that they did not know that what they were doing was unethical. Consequently, the OHRP has now mandated that all researchers (which includes everyone listed on the research protocol) must have training in the protection of human subjects in order for research to proceed. It is up to the IRB to document that all the researchers have training. Excellent online training courses can be found on the websites of the National Institutes of Health's Office of Human Subjects Research (OHSR) (http://ohsr.od.nih.gov/) and at the CITI Course in the Protection of Human Research Subjects (http://www.citiprogram.org/).

Final thoughts: consequences of conducting unreviewed research

Always get your research reviewed by the IRB and follow your procedures as you have outlined them. It is worth knowing what the consequences to the institution and the researcher are if something untoward happens to a subject in a study that was conducted without IRB review. The institution (university, hospital, etc.) has civil as well as criminal liability. It will also lose all federal funding, regardless of whether it is research related or not. The investigative team (the researchers) also has civil as well as criminal liability (meaning that they can be sued and go to jail). They can also lose the ability to apply for future research funding, and they also can face the loss of their job, even if they have tenure. The bottom line is that if you are going to conduct stress research in the field, it should be submitted to and approved by the IRB.

References

Federal Register (1998). *Protection of Human Subjects: Categories of Research That May Be Reviewed by the Institutional Review Board (IRB) Through an Expedited Review Procedure*, 63 FR 60364–60367, November 9, 1998.

Title 45 CFR Part 46 (2005). *Protection Of Human Subjects*, Department of Health and Human Services, National Institutes of Health Office for Protection from Research Risks, revised June 23, 2005.

OHRP IRB Handbook (N.D.).http://www.hhs.gov/ohrp/irb/irbguidebook.htm

Title 45 CFR Parts 160, 164 (2002). *Privacy of Individually Indentifiable Health Information*, revised August 14, 2002.

Vanderpool, H. Y. (1996). Introduction and overview: ethics, historical case studies and the research enterprise. In *The Ethics of Research Involving Human Subjects: Facing the 21st Century*, ed. H. Y. Vanderpool. Frederick, MD: University Publishing Group, pp. 1–30.

11 *Epilog: summary and future directions*

GARY D. JAMES AND GILLIAN H. ICE

Despite the fact that the biological and behavioral responses to the ever-changing conditions in real life define a very important and central component of survival, it is interesting that very few human biological and anthropological researchers include measurements of these responses in their evolutionary study of contemporary humans. The psychobiology of stress, in fact, is often overlooked in evolutionary and ecological studies of human population variation. Part of this oversight may be related to a lack of information on how to study stress under the conditions of real life. This volume will hopefully provide some guidance for researchers interested in expanding their field research to include an assessment of the adaptation to stress.

It is apparent from the contributions to this volume that there are both conceptual and technical issues to consider when evaluating psycho-biological adaptation in field studies. As articulated in Chapter 9, researchers need to formulate a plan, and as part of that formulation, carefully determine what is or is not "stressful" in the context of their study. Perhaps a better way of framing this formulation is that the researcher must consider what in the human ecology induces or will induce psychobiological change in their subjects. Chapters 2 through 7 detail the present state of knowledge and current methods of measurement of the cognitive, endocrinological and physiological responses to stress. Each measure has its limitations and many present challenges for field research that are often unimportant in laboratory or clinical settings. Which measures will be used is often a practical decision that is based on technical constraints and population characteristics, as well as the research question. As noted in Chapter 2, regardless of measures chosen, it is clear that the researcher must pay careful attention to demographic, behavioral and social characteristics of the sample as they often affect results and conclusions.

There are still few measures that are truly field-friendly. While many can be easily used in a field setting in the developed world, with easy

266

access to refrigeration and laboratories, many methods are challenging to use in remote areas with limitations in electricity and laboratory space. We recommend that researchers conduct pilot tests of methods because one often does not know the particular limitations in a field site. For example, Ice and Yogo (2005) attempted to use the saliva collection method used by Flinn and England (1995; 1997) in their recent research on stress in Kenya. Flinn collects saliva at specific times by walking around town and gathering subjects. In a pilot study, Ice attempted to drive from homestead to homestead in a rural area of western Kenya. It quickly became clear that the homesteads were too far apart to collect saliva in a timely manner. The method had to be adapted for the larger project. Without a pilot study, this methodological challenge could have seriously hindered the project.

Clearly, stress research has been largely conducted in modernized societies, mostly in the West. Far more research needs to be conducted in other settings to better our knowledge of how psychosocial stress impacts human biological variation. However, cross-cultural studies present methodological challenges. Where does the methodology of measuring stress go from here? The answer to that question is closely tied to what kinds of studies need to be done in the future, and whether it is important to be able to compare stress assessments across populations. If comparability is a goal for stress response studies, then similar research questions need to be posed and addressed in many different populations, or at least the research protocols should be compatible with broader comparisons with other studies. Because stress responses are so intimately tied to the ecological constraints specific to populations, broad standardization of protocols may not be possible. Technically, however, in many of the stress measures evaluated, it is possible to standardize at least the sampling frame (e.g. taking measurements every 15 minutes) or means of measurement expression (e.g. urinary rates vs. concentrations or ratios).

We would also recommend that future studies incorporate several stress measures, not just one. Stress studies still largely use single stress markers. As a result, we have little knowledge of the variable response of stress markers to various stressors. We also know little about how the stress markers interact with each other. Research using multiple markers will further our understanding of the performance of individual markers as well as increase our understanding of the entire stress process. The cognitive and physiological processes that can be measured with stress are integrated, and it is likely that broader systemic assessments will provide greater insight into the adaptive and morbid outcomes of stress.

Currently, the majority of stress studies are cross-sectional in nature. This methodological approach has limited our ability to connect stressors and the stress process to disease outcomes. Instead, researchers have relied on a chain of inference. In other words, we have demonstrated that stress has an impact on various biomarkers of health but rarely demonstrated that stress has an impact on morbidity or mortality. For example, we know that psychosocial stress increases blood pressure. But how much exposure is needed to increase the risk of hypertension or myocardial infarction? The answer to this question can only be answered through longitudinal study and analysis.

Finally, in many ways, the study of stress and adaptation in human population biology is still in its infancy. This volume has attempted to outline current field methods of making measurements in those physiological and behavioral systems that characterize the human responses to stress. There have been remarkable advances in assay and monitoring technology as well as methodological and theoretical constructs in the study of cognitive, hormonal, cardiovascular and immune measurements. While there remain limitations with regard to making accurate assessments of the human response to real life stressors, as research interest increases, methodology will be likely to continue to improve.

References

Flinn, M. and England, B. (1995). Family environment and childhood stress. *Current Anthropology*, **36**, 854–66.

  (1997). Social economics of childhood glucocorticoid stress response and health. *American Journal of Physical Anthropology*, **102**, 33–54.

Ice, G. H. and Yogo, J. (2005). Development of a culturally specific scale to measure perceived stress: The Luo Perceived Stress Scale. *Field Methods*, **17**(4), 394–411.

Index